Social Dynamics

Social Dynamics

Brian Skyrms

OXFORD
UNIVERSITY PRESS

OXFORD
UNIVERSITY PRESS

Great Clarendon Street, Oxford, OX2 6DP,
United Kingdom

Oxford University Press is a department of the University of Oxford.
It furthers the University's objective of excellence in research, scholarship,
and education by publishing worldwide. Oxford is a registered trade mark of
Oxford University Press in the UK and in certain other countries

First Edition published in 2014

Impression: 1

Published in the United States of America by Oxford University Press
198 Madison Avenue, New York, NY 10016, United States of America

British Library Cataloguing in Publication Data

Data available

Library of Congress Control Number: 2014933966

ISBN 978-0-19-965282-2 (hbk.)
ISBN 978-0-19-965283-9 (pbk.)

Printed in Great Britain by
CPI Group (UK) Ltd, Croydon, CR0 4YY

Contents

List of Figures

List of Tables

Acknowledgments

The chapters collected here are reprinted with the kind permission of the original publishers:

Evolution and the Social Contract, in *The Tanner Lectures on Human Values* 28. Salt Lake City: University of Utah Press, pp. 47–69. 2009.

Trust, Risk, and the Social Contract, *Synthese* 160: 21–5. © Springer 2008.

Bargaining with Neighbors: Is Justice Contagious? *Journal of Philosophy* 96: 588–98. © *The Journal of Philosophy*, Inc. 1999.

Stability and Explanatory Significance of Some Simple Evolutionary Models, *Philosophy of Science* 67: 94–113. © the Philosophy of Science Association 2000.

Dynamics of Conformist Bias, *The Monist* 88: 260–9. © *The Monist* 2005.

Chaos and the Explanatory Significance of Equilibrium: Strange Attractors in Evolutionary Game Dynamics, *PSA* 1992, 2: 374–94. © the Philosophy of Science Association 1993.

Evolutionary Dynamics of Collective Action in N-person Stag Hunt Dilemmas, *Proceedings of the Royal Society* B. 276: 315–21. © The Royal Society 2008.

Learning to Take Turns, *Erkenntnis* 59: 311–48. © Springer 2003.

Evolutionary Considerations in the Framing of Social Norms, *Philosophy, Politics and Economics* 9: 265–273. © Sage Publications 2010.

Learning to Network, in *The Place of Probability on Science*, ed. E. Eells and J. Fetzer. Springer, 277–87. © Springer 2004.

A Dynamic Model of Social Network Formation, *Proceedings of the National Academy of Sciences of the U.S.A.* 97: 9340–6, 2000.

Network Formation by Reinforcement Learning: The Long and the Medium Run, *Mathematical Social Sciences* 48: 315–27. © Elsevier B.V. 2004.

Time to Absorption in Discounted Reinforcement Models, *Stochastic Processes and their Applications* 109: 1–12. © Elsevier B.V. 2003.

Learning to Signal: Analysis of a Micro-level Reinforcement Model, *Stochastic Processes and their Applications* 119: 373–419. © Elsevier B.V. 2008.

Inventing New Signals, *Dynamic Games and Applications* 2: 129–45. © Springer 2011.

Signals, Evolution, and the Explanatory Power of Transient Information, *Philosophy of Science* 69: 407–28. © the Philosophy of Science Association 2002.

Co-Evolution of Pre-play Signaling and Cooperation, *Journal of Theoretical Biology* 174: 30–5. © Elsevier Ltd. 2011.

Evolution of Signaling Systems with Multiple Senders and Receivers, *Philosophical Transactions of the Royal Society* B 364: 771–9. © The Royal Society 2008.

Introduction

Social Dynamics could mean all sorts of things. In the essays in this book it means analysis of adaptive dynamics in prototypical models of social interaction.

By adaptive dynamics, I simply mean a dynamics that moves in the direction of that which succeeds, or seems to succeed, better that the alternatives. Inspiration comes from evolutionary dynamics. There are versions for both large and small populations, with and without mutation. Evolution may be cultural rather than biological, with imitation rather than replication driving the dynamics. Individuals who repeatedly interact may adapt to one another's actions by various more or less sophisticated kinds of individual learning. These are all instances of adaptive dynamics.

In contexts of strategic interaction, everyone aiming for the best may very well lead to the worst. Or it may lead nowhere definite at all, as is the case when the dynamics is cyclic, or even chaotic. Adaptive dynamics need not lead to adaptation. Taking a shortcut by assuming that it does may lead to an erroneous analysis.

The interactions discussed here are simple. The aim is to isolate and study essential aspects of social dynamics. The drawback of this approach is that the real world is always much more complex than the models. The advantage is that one can really analyze the dynamics of the models, sometimes with really surprising results. If the interactions are well chosen, they can, when understood, serve as building blocks for more complicated models. I have tried to move beyond the ubiquitous Prisoner's Dilemma to also include signaling, bargaining, multiplayer Stag Hunt, division of labor, dynamic network formation, and combinations of these. Even the simplest interactions have delivered surprises when subjected to rigorous analysis.

I was led to social dynamics by a winding road. First there was the question of decision instability in rather esoteric situations in which one's coming to a decision may generate information about the world relevant to that decision. It was raised by Allan Gibbard and Bill Harper in a discussion of "The Man who Met Death on the way to Damascus."

This led to a book on dynamics of rational deliberation in which players in interactive decision problems try to reason their way to equilibrium. Each has a model of the other's reasoning and runs through a fictitious back and forth adaptive process to see where it leads. Their models of each other need not be at all accurate, so all sorts of things may happen. That led to adaptive dynamics that isn't fictitious but actual—the subject of this book.

Many of the essays collected here are collaborative, and the heavy lifting is often done by my co-authors. There is no grand overarching theory, but rather an ongoing program of research.

References

Gibbard, A. and W. Harper (1981) "Counterfactuals and Two Kinds of Expected Utility." In *IFs* ed. Harper et al. Dordrecht: Reidel.

Skyrms, B. (1982) "Causal Decision Theory." *Journal of Philosophy* 79: 695–711.

Skyrms, B. (1990) *The Dynamics of Rational Deliberation.* Cambridge, MA: Harvard University Press.

PART I

Correlation and the Social Contract

Introduction

I have chosen my Tanner lecture, *Evolution and the Social Contract*, to begin this book for two reasons. The first is that it gives a broad overview of many of the issues that are discussed in greater detail later in the book. The second is that it focuses attention on correlation mechanisms as a key to understanding the social contract.

Limiting oneself to simple tractable models does not mean limiting oneself to one model of the social contract, as some thinkers seem to do. In particular, I do not think that the Prisoner's Dilemma and its many-person generalizations are the unique key to understanding cooperation. Many situations discussed as Prisoner's Dilemma are really Stag Hunt games. Many supposed explanations of cooperation in Prisoner's Dilemma are really ways of transforming the game into a Stag Hunt. Ken Binmore has showed the importance of bargaining games to social contract theory. (See also Richard Braithwaite's inaugural lecture.) These games combine—Stag Hunt to get the goods, Bargaining to divide the spoils. In some situations efficient Stag Hunting may require specialized roles. This introduces Division of Labor games. David Lewis focused attention on Signaling games. Signaling may be important for coordinating action. Individuals often interact in a social network, which itself evolves.

Social structure should be analyzed in terms of correlation. The importance of correlation in evolution was pointed out by William Hamilton in a seminal series of papers, starting in 1964. Correlation impacts every-thing in the analysis of social behavior. Basically, in the presence of an exogenous correlation mechanism, the results of rational choice game theory are out the window. Hamilton uses correlation to explain the existence of both altruism and spite. The correlation devices found in human society appear, at least from our perspective, far richer than those found throughout the animal world. Social institutions are all correlation devices, controlling the large-scale interaction structure of societies. Social dynamics should focus on social interactions in the presence of correl-ation, and on the dynamics of correlation devices themselves.

References

Binmore, K. (1994) *Game Theory and the Social Contract I: Playing Fair.* Cambridge, MA: MIT Press.

Binmore, K. (1998) *Game Theory and the Social Contract II: Just Playing.* Cambridge, MA: MIT Press.

Braithwaite, R. (1955) *The Theory of Games as a Tool for the Moral Philosopher.* Cambridge: Cambridge University Press.

Hamilton, W. (1964) "The Genetical Evolution of Social Behavior I and II." *Journal of Theoretical Biology* 7: 1–52.

Lewis, D. (1969) *Convention.* Cambridge, MA: Harvard University Press.

1

Evolution and the Social Contract

1. Dewey and Darwin

Almost one hundred years ago John Dewey wrote an essay titled "The Influence of Darwin on Philosophy." At that time, he believed that it was really too early to tell what the influence of Darwin would be: "The exact bearings upon philosophy of the new logical outlook are, of course, as yet, uncertain and inchoate." But he was sure that it would not be in providing new answers to traditional philosophical questions. Rather, it would raise new questions and open up new lines of thought. Toward the old questions of philosophy, Dewey took a radical stance: "Old questions are solved by disappearing . . . while new questions . . . take their place."

I don't claim that the old philosophical questions will disappear, but my focus is the new ones. Evolutionary analysis of the social contract—be it cultural or biological evolution—*does not tell you what to do*. Rather, it attempts to investigate how social conventions and norms evolve—how social contracts that we observe could have evolved and what alternative contracts are possible.

The tools for a Darwinian analysis of the social contract are those of evolutionary game theory. From the theory of games it takes the use of simple stylized models of crucial aspects of human interaction; from evolution it takes the use of adaptive dynamics. The dynamics need not have its basis in genetics; it may as well be a dynamic model of cultural evolution or of social learning (Weibull 1995; Björnerstedt and Weibull 1996; Samuelson 1997; Schlag 1998). No part of what follows implies any kind of genetic determinism or innateness hypothesis. Problems of cooperation may be solved by genetic evolution in some species and by cultural evolution in others.

Here are three features of the evolutionary approach to bear in mind in the ensuing discussion:

1. Different social contracts have evolved in different circumstances.
2. Existing social contracts are not altogether admirable.
3. We can try to change the social contract.

Correlation and the evolution of cooperation

Cooperation may be easy or hard, or somewhere in between. Here is a game-theory model of an easy problem of cooperation (Binmore 2005) Prisoner's Delight:

	PRISONER'S DELIGHT	
	Cooperate	Defect
Cooperate	3	1
Defect	2	0

The entries in the table represent the payoffs of Row's strategy when played against Column's strategy. If your partner cooperates, you are better off if you do as well, for a payoff of 3 rather than 2. If your partner defects, you are still better off cooperating for 1 rather than 0, although your partner does even better with 2. So no matter what your partner does, you are better off cooperating. Your partner is in the same situation, and reasons likewise. It is easy to cooperate.

We can change the story a little bit to fit a different kind of situation. Perhaps if your partner defects, your attempt at cooperation is counterproductive. You are better off defecting. This change gives us the game known as the Stag Hunt:

	STAG HUNT	
	Cooperate	Defect
Cooperate	3	1
Defect	2	2

In the Stag Hunt, what is best for you depends on what your partner does. If you both cooperate, you are both doing the best thing given your

partner's action—likewise if you both defect. Cooperation is more difficult. It is an equilibrium, but not the only one.

Another modification of the situation calls for a different game.
Suppose that defecting against a cooperator actually pays off. Then we
have the Prisoner's Dilemma:

	PRISONER'S DILEMMA	
	Cooperate	Defect
Cooperate	3	1
Defect	4	2

Your optimal act again no longer depends on what your partner does.
Now it is to defect. Cooperation is hard.

Each of these games is a reasonable model of some social interactions.
Two men sit in a rowboat, one behind the other. Each has a set of oars.
They have been out fishing, and a hot dinner awaits them across the lake.
If one doesn't row for some reason, the other will row to get them there;
if one does row, the other prefers to row also to get home faster. This is
Prisoner's Delight. Cooperation is easy. Now change the picture. They sit
side by side, and each has one oar. One man rowing alone just makes the
boat go in circles. This is a *Stag Hunt.*[1] Back to the first rowboat with two
sets of oars, but take away the hot dinner on the opposite shore and
suppose that the men are very tired. They could camp on this shore for
the night, although they would prefer to get back to the opposite shore.
But either prefers not to row no matter what the other does. This is a
Prisoner's Dilemma.

Consider any reasonable adaptive dynamics operating on a large
population of individuals paired at random to play one of these games.
Then in Prisoner's Delight cooperation takes over the population, in
Prisoner's Dilemma defection goes to fixation, and in Stag Hunt one or
the other may prevail depending on the initial composition of the
population.

There is an enormous literature devoted to explaining the evolution of
cooperation in the Prisoner's Dilemma, while the other two games are

---

[1] This is David Hume's famous example from the *Treatise of Human Nature.* For more
game theory in Hume, see Vanderschraaf 1998.

relatively neglected. Everyone wants to crack the hardest problem—the *evolution of altruism*.[2]

All these accounts of cooperation in Prisoner's Dilemma either (1) use an interaction that is not really a Prisoner's Dilemma or (2) use pairings to play the game that are not random. It must be so.[3] Suppose the interaction is a Prisoner's Dilemma, and pairings are random in a large population. Then cooperators and defectors each have the same proportion of partners of each type. Defectors must on average do better than cooperators. Replicator dynamics increases the proportion of the population of the type that does better, and that's all there is to it.

But if nature somehow arranges for positive correlation—for cooperators to meet cooperators and defectors to meet defectors more often than they would with random matching—then it is possible for cooperators to do better than defectors. The point is obvious if we consider perfect correlation. Then the relevant comparison in payoffs is not vertical but diagonal:

	PRISONER'S DILEMMA	
	Cooperate	Defect
Cooperate	3	1
Defect	4	2

Every explanation of the evolution of cooperation in real Prisoner's Dilemmas—kin selection, group selection, repeated games, spatial interaction, static and dynamic interaction networks, and all the rest—works by providing a mechanism that induces correlation in plays of the Prisoner's Dilemma. This was clear to William Hamilton and to George Price back in the 1960s (Hamilton 1964, 1995; Price 1970; Eshel and Cavalli-Sforza 1982; Frank 1995). (A version of Hamilton's rule for kin selection can be derived just from the positive correlation.)

But to say that a mechanism can sometimes generate enough positive correlation to maintain cooperation in the Prisoner's Dilemma is not to

---

[2] Binmore (1994) devotes a chapter titled "Squaring the Circle in the Social Sciences" to attempt to justify cooperation in the one-shot Prisoner's Dilemma.

[3] This is the "Iron Rule of Selfishness" of Bergstrom 2002. See also Eshel and Cavalli-Sforza 1982.

say that it can always do so. In some circumstances correlation may fall short. Thus, in each of these accounts, an examination of specific correlation mechanisms is of interest. And often a scenario can be analyzed *both* as Prisoner's Dilemma with correlation *and* as a larger game in which the plays of Prisoner's Dilemma are embedded. I will illustrate with two examples.

Axelrod (1984) directs our attention to the *shadow of the future*. Cooperation may be maintained not by immediate payoffs but by the consequences of current actions on future cooperative behavior of partners. In this he follows Thomas Hobbes and David Hume:

HOBBES: He, therefore, that breaketh his Covenant, and consequently declareth that he think that he may with reason do so, cannot be received into any society that unite themselves for Peace and Defense, but by the error of them that receive him.

HUME: Hence I learn to do a service to another, without bearing him any real kindness; because I foresee, that he will return my service, in expectation of another of the same kind, and in order to maintain the same correspondence of good offices with me and with others.

Axelrod, following John Nash and other founding fathers of game theory,[4] analyzes the shadow of the future using the theory of indefinitely repeated games. Suppose that the probability that the Prisoner's Dilemma will be repeated another time is constant. This (somewhat far-fetched) idealization of geometrical discounting of the future allows us to sum an infinite series and compute the expected payoffs of strategies in the large repeated game. For simplicity, we consider only two strategies in the repeated game, *Always Defect* and *Tit for Tat*. A Tit for Tat player initially cooperates and then does what was done to him in the preceding round, and Always Defect is self-explanatory. Two players remain matched for the whole repeated game, but this restrictive assumption can be relaxed in more complicated "community enforcement" models (Sugden 1986; Milgrom et al. 1990; Kandori 1992; Nowak and Sigmund 1998).

Since the payoffs on each individual play are those of Prisoner's Dilemma, the strategies in the repeated games must induce a correlation

[4] The use of discounted repeated games to explain cooperation in Prisoner's Dilemma is already to be found in Luce and Raiffa 1957, with no claim of originality. John Nash is reported to have invented the explanation in conversation.

between individual plays of Cooperate and Defect if cooperation is not to be driven to extinction. The presence of Tit for Tat players in the population is this correlation device. They *always* cooperate with each other and quickly learn to defect against defectors.

What is the larger game in which plays of Prisoner's Dilemma are embedded? Using the version of Prisoner's Dilemma given before and probability of another trial as six-tenths, we get:

	Tit for Tat	Always Defect
Tit for Tat	7.5	4
All Defect	7	5

This is a Stag Hunt. There are two stable equilibria, one where everyone always plays Tit for Tat and one where everyone always plays Always Defect. Which one you get depends on initial population proportions. In our example an initial population split equally evolves to universal defection.

Sober and Wilson (1998) direct our attention to group selection. Consider the haystack model of Maynard Smith (1964). In the fall, farmers cut hay and field mice randomly colonize the haystacks. In the haystacks they play the Prisoner's Dilemma and reproduce according to the payoffs. In the spring the haystacks are torn down, the mice scatter, and the cycle is continued. If there are enough generations of mice in the life of a haystack, then it is possible for cooperators to do better on average than defectors. This is because differential reproduction within haystacks creates positive correlation in the population. In haystacks colonized by cooperators and defectors, defectors take over. Then cooperative haystacks out-reproduce noncooperative ones.

What is the larger game within which the plays of the Prisoner's Dilemma are embedded? Following Ted Bergstrom (2002), we consider the game played by founding members of a haystack. Their payoff is the number of descendants in the spring, at the end of the life of a haystack. Analysis of this founders game shows that it too is a Stag Hunt. There are two equilibria, one with all cooperators and one with all defectors. Which you get depends on where you start.

Each of these models explains how a population of cooperators can be at a stable equilibrium, but neither explains the origin of cooperation.

That is because noncooperation is also a stable state, and the transition from noncooperation to cooperation is left a mystery.

We are left with the question of how it is possible to evolve from the noncooperative equilibrium to the cooperative one in interactions with the structure of the Stag Hunt. Axelrod and Hamilton (1981) raise this question and take kin selection and cooperation in family groups as an origin of cooperative behavior. Beyond this there are a few other good answers available. I will focus on three of them.

The first, due to Arthur Robson (1990), is the use of a signal as a *secret handshake*. Consider a population of defectors in the Stag Hunt game. Suppose that a mutant (or innovator) arises who can send a signal, cooperate with those who send the same signal, and defect against those who do not. The new type behaves like a native against the natives and like a cooperator against itself, and so can (slowly) invade. (The signal need not really be a secret, but it should not be in current use by defectors for other purposes.) Once cooperators are established, it does not matter if the signaling system somehow falls apart, because in Stag Hunt—unlike Prisoner's Dilemma—no one can do better against a population of cooperators.

The second involves a special kind of local interaction with neighbors. Instead of the random encounters of the usual model, individuals interact with their neighbors on a spatial grid (or some other spatial structure).[5] They play a Stag Hunt game with each neighbor, and cultural evolution proceeds by imitation of the most successful (on average) strategy in the neighborhood. For biological evolution, there is an alternative interpretation in which success translates into reproduction in the neighborhood. We thus have both neighborhood interaction and neighborhood imitation.

Eshel et al. (1999) point out that these two neighborhoods need not be the same. For a biological example, take a plant that interacts locally but disperses its seeds widely. On the cultural side we can consider cases where the flow of information allows an individual to observe success

---

[5] There are pioneering papers by Pollock (1989), Nowak and May (1992), and Hegselmann (1996). These are models where the interaction is Prisoner's Dilemma, or in the case of Nowak and May at a bifurcation between Prisoner's Dilemma and Hawk-Dove. Ellison (1993, 2000) discusses local interaction models of the Stag Hunt in which the <Defect, Defect> equilibrium is risk-dominant. In his models it is the noncooperators who can invade and quickly take over. The differences between Ellison's models and ones in which cooperators can invade and take over are discussed in Skyrms (2004).

beyond the confines of immediate interactions. This is the crucial modification to local interaction models that allows robust evolution of cooperation: the imitation neighborhood must be sufficiently larger than the interaction neighborhood. Then a small clump of contiguous cooperators will grow and eventually take over the population. That is because defectors can see the success of the internal cooperators who interact only with cooperators, and imitate them.

Kevin Zollman (2005) shows that the secret handshake and local interaction work especially well together. The signal used as a secret handshake now needs only to be a *local* secret to work. It could be used elsewhere in the population to mean all sorts of other things. This makes the hypothesis of the existence of an unused signal much more plausible. The local secret handshake can then facilitate the initial formation of a large-enough clump of contiguous cooperators to allow cooperation to spread by imitation.

The third answer involves dynamic networks. Instead of constraining individuals to interact with neighbors on a fixed structure, we can allow the interaction structure to evolve as a result of individuals' choices (Skyrms and Pemantle 2000 (Chapter 11 in the present volume); Bonacich and Liggett 2003; Liggett and Rolles 2004; Pemantle and Skyrms 2004a (Chapter 12), 2004b (Chapter 13); Skyrms and Pemantle 2004 (Chapter 10); Pacheco et al. 2006; Santos et al. 2006; Skyrms 2007). Cooperators want to interact with each other. In Stag Hunt—unlike Prisoner's Dilemma—noncooperators do not much care. Cooperators and defectors may not wear their strategies on their lapels, but even if it is not so easy to tell cooperators from noncooperators, it is not hard to learn. Robin Pemantle and I show how even naive reinforcement learners will form cooperative associations in Stag Hunt provided the social-network structure is sufficiently fluid (Skyrms and Pemantle 2000; Pemantle and Skyrms 2004a, 2004b). The conclusion is robust to various variations in the learning dynamics (Skyrms 2004, 2007). These cooperative associations could then be the focal points for imitation to spread cooperation, as in the foregoing discussion.[6]

Theory is borne out in laboratory studies of human interactions. There is experimental evidence that there are many different types of individuals

---

[6] For other models of correlation induced by partner choice, see Wright 1921, where the literature begins; Hamilton 1971; Feldman and Thomas 1987; Kitcher 1993; Oechssler 1997; Dieckmann 1999; and Ely 2002.

(Burlando and Guala 2005; Page et al. 2005; Fischbacher and Gächter 2006) and that given the opportunity and the requisite information, cooperators will learn to associate with one another to their own benefit.[7]

It is evident that the three preceding accounts of transition from the noncooperative equilibrium of the Stag Hunt to the cooperative one also rely on the establishment of correlated interactions. They share two distinctive features that are missing in many accounts of cooperation:

1. Sufficient positive correlation can be established by a few cooperators in a large population of noncooperators.
2. Once the cooperative equilibrium is reached, it can be maintained even if the correlation fades away.

Negative correlation and spite

If social structure can create positive correlation of encounters, it can also create negative correlation. Negative correlation can overturn the conclusions of conventional game theory just as radically as positive correlation.

	PRISONER'S DILEMMA	
	Cooperate	Defect
Cooperate	3	1
Defect	4	2

	STAG HUNT	
	Cooperate	Defect
Cooperate	3	1
Defect	2	2

	PRISONER'S DELIGHT	
	Cooperate	Defect
Cooperate	3	1
Defect	2	0

---

[7] For an experiment in a public goods–provision game with voluntary association, see Page et al. 2005.

Consider the effect of perfect negative correlation—of always meeting the other type—on our three games. We are now comparing the other diagonal: Not only is defection restored in the Prisoner's Dilemma and favored in the Stag Hunt, but it is also imposed on Prisoner's Delight. In this last case an individual hurts himself by defecting, but hurts his partner more. This is a case of spiteful behavior. Hamilton and Price also showed how negative correlation is the key to the evolution of spite.

Both spite and altruism on their face appear to violate the rational choice paradigm. Both have an evolutionary explanation in terms of correlated interactions. Yet spite rarely receives the attention given to altruism. A search on Google Scholar for "Evolution of Altruism" gets 1,570 hits, while one for "Evolution of Spite" gets 32. One cannot help asking whether this is due to a Pollyanna bias. It's enjoyable to write about the sunny side of human nature. But the world is full of spiteful behavior: feuds, vendettas, and senseless wars. It is as important to study it as to study altruism.

The way to study spite is to study endogenous correlation mechanisms. In some kinds of repeated interactions, the shadow of the future may sustain spite. Johnstone and Bshary (2004) recently analyzed the persistence of spite in a repeated-game setting. In repeated contests, a reputation for fighting too hard—to one's own detriment as well as the greater detriment of the opponent—may enable one to win future contests more easily. The application need not be restricted to animal contests.

It might be remarked that this is not really spite in the larger game but only self-interest, just as Hobbes claimed that cooperative behavior in a repeated Prisoner's Dilemma may just be self-interest in the longer view. It is useful to be able to look at the phenomenon from both perspectives.

Successful invasion of a spiteful type into a nonspiteful population can be sustained by local interaction. A strain of E. coli bacteria produces, at some reproductive cost to itself, a poison that kills other strains of E. coli—but to which it is impervious. It cannot invade a large random mixing population because a few poisoners cannot do that much damage to the natives and the natives out-reproduce the poisoners. But in a spatial, local interaction setting, a clump of poisoners can take over. These phenomena have been observed in the laboratory, with the random encounters taking place in a well-stirred beaker and local interactions taking place in a petri dish. Theoretical analysis by Durrett and Levin (1994) and Iwasa et al. (1998) is a complement to the local interaction model of evolution of cooperation. If we put this case in the framework of Eshel, Shaked, and Sansone (1999) of the last section, we find that a large interaction neighborhood and a small imitation neighborhood robustly favor the evolution of spite (see Skyrms 2004).

Group-selection models are not without their spiteful aspects. Suppose that the farmer never tears down the haystacks—the population islands that they represent remain isolated. Then for a mouse, its haystack becomes its world, and this makes all the difference (see Gardner and West 2004). Its haystack's population becomes a population unto itself, within which evolution takes place. The carrying capacity within a haystack is limited, so we are dealing with a small, finite population. This, in itself, induces negative correlation even if pairs of individuals form at random within the haystack, because an individual does not interact with herself. This effect is negligible in a large population but can be significant in a small population. (For a transparent example, consider a population consisting of four individuals, two C and two D. Population frequencies are 50–50, but each type has probability ⅔ of meeting the other type and ⅓ of meeting its own.)

If a defector is introduced into a haystack full of cooperators—a mutant or a migrant—he can cause problems. If the interaction is Prisoner's Dilemma, defectors will, of course, take over. But in small populations, with some versions of Stag Hunt and even Prisoner's Delight, defectors can still invade as a result of negative correlation.

For any positive value of e, the following is a version of Prisoner's Delight—an individual prefers to cooperate no matter what the other does:

	PRISONER'S MILD DELIGHT	
	Cooperate	Defect
Cooperate	$2 + e$	e
Defect	2	0

For any finite population, there is an e—yielding some version of Prisoner's Mild Delight—such that a spiteful defector can invade.[8]

---

[8] For example, suppose 1 defector is introduced into a population of N cooperators. Individuals pair at random. Since the defector cannot interact with himself, he always pairs with a cooperator for a payoff of 2. Cooperators pair with the defector with probability $(1/N)$ and with other cooperators with probability $(N-1)/N$, for an average payoff of $[(N-1)/N]*2 + e$. So if $e < (2/N)$, a spiteful mutant does better than the native cooperators.

These three examples serve to indicate that the evolution of spite is an aspect of the evolution of the social contract that is worthy of more detailed study. There is no reason to believe that they exhaust the mechanisms for negative correlation that may be important in social interactions.

To stop here would be to represent the social contract as a neat and simple package of problems. But the social contract is not neat and simple.

Bargaining

Prisoner's Delight, Stag Hunt, and Prisoner's Dilemma are not the only games that raise issues central to the social contract. We could separate the issues of cooperating to produce a public good and deciding how that good is to be divided. This is the philosopher's problem of distributive justice, and it brings bargaining games to center stage (Braithwaite 1955; Rawls 1957; Gauthier 1985, 1986; Sugden 1986; Binmore 1994, 1998, 2005).

Consider the simplest Nash Bargaining game. Two players have bottom-line demands for a share of a common good. If the demands exceed the available good, agreement is impossible and players get nothing. Otherwise, they get what they demand. We simplify radically by assuming that there are only three possible demands: one-third, one-half, and two-thirds. Evolutionary dynamics in a large random-encounter environment, with and without persistent random shocks, is well studied.

Allowing differential reproduction to carry a population to equilibrium, there are two possibilities. The population may reach an egalitarian consensus, where all demand one-half. Or it may reach a polymorphic state, where half the population demands two-thirds and half the population demands one-third (Sugden 1986). Greedy players get their two-thirds half the time; modest players get their one-third all the time. This inefficient polymorphism wastes resources, but it is evolutionarily stable and has a significant basin of attraction. Persistent shocks can allow a population to escape from this polymorphic trap and favor the egalitarian norm in the very long run, but the inefficient polymorphism remains a possibility for medium-run behavior.[9]

---

[9] Most of what we know about this is due to Peyton Young. See Young 1993a, 1993b, 1998; and Binmore et al. 2003.

However, the effect of correlation mechanisms is rarely discussed in connection with Nash Bargaining. If correlation plays an important role in producing a surplus to be divided, might it not also play an important role in deciding how the division takes place? Positive correlation of demand types obviously favors the egalitarian solution. Those who ask for equal shares do best when they meet each other. Negative correlation is more complicated. Greedy players who demand two-thirds do very well if paired with modest players who ask for one-third, but not well if paired with those who ask for half. If negative correlation initially allows greedy players to out-reproduce all others, it cannot be maintained because they run out of modest players. But a greedy-modest polymorphism is a real possibility if the negative correlation is of the kind that sufficiently disadvantages the egalitarians. Possibilities are multiplied if we allow more demand types. It should be of interest to look at specific correlation mechanisms.

If we allow individuals to bargain with neighbors on a spatial grid, islands of egalitarianism spontaneously form. This generates positive correlation, and if individuals emulate their most prosperous neighbors, egalitarianism takes over the population. This is a very rapid process, as other types who interact along the edges of the egalitarian islands are quickly converted to asking for half (J. Alexander and Skyrms 1999 (Chapter 3); J. Alexander 2007; Skyrms 2004). Equal sharing is contagious.

If we add costless signaling to a large-population random-encounter evolutionary model of bargaining, complicated correlations arise and then fade away. Cooperators establish a positive correlation with cooperators. Greedy types establish a negative correlation with themselves. Although these correlations are transient, their effect is that the basin of attraction of the egalitarian equilibrium is greatly enlarged (see Skyrms 2004).

Axtell et al. (2006) investigate a related model where individuals have one of two "tags" and can condition their action on the tag of their partner in a bargaining game. But there is a different dynamics. Instead of evolution by replication or imitation, they consider a rational-choice model. Things may fall out in various ways—here is one. When interacting with those having the same tag, individuals share alike. But in interactions between tags, one tag type becomes greedy and always demands two-thirds and the other becomes modest and always demands one-third. In this equilibrium tags are used to set up both positive and negative correlations between behaviors. Both correlations are perfect: demand-half behaviors always meet themselves, and the other two demand behaviors always meet each other. The result is egalitarianism within tag types and inegalitarian distribution between types. Axtell et al. see this as a spontaneous emergence of social classes.

We can also see the spontaneous emergence of social classes in a dynamic social network (Skyrms 2004). The classes are stabilized by rational choice but destabilized by imitation. Depending on the details and timing of the dynamics, the social network may end up egalitarian or with a class structure.

Division of Labor

So far, negative correlation has played a rather sinister role in this story. It is not always so. In cooperating to produce a common good, organisms sometimes discover the efficiency of division of labor, and find a way to implement it. Modern human societies are marvels in their implementation of division of labor; so are the societies of cells in any multicellular organism. On the most elementary level, we can suppose that there are two kinds of specialists that an individual can become, A and B, and that these specialists are complementary. On the other hand, an individual might not specialize at all but rather, less efficiently, do what both specialists do. This gives us a little division-of-labor game:[10]

	DIVISION OF LABOR		
	Specialize A	Specialize B	Go It Alone
Specialize A	0	2	0
Specialize B	2	0	0
Go It Alone	1	1	1

In a random-encounter setting, specialists do badly. Positive correlation makes it worse. What is required to get division of labor off the ground is

---

[10] For analysis of different division-of-labor games, motivated by evolution of coviruses, see Wahl 2002.

the right kind of negative correlation. Not all the correlation mechanisms that we have discussed here do the trick.[11] What works best is dynamic social-network formation, where the network structure evolves quickly. Specialists quickly learn to associate with the complementary specialists, and then specialists outperform those who go it alone. The effect of correlation depends on the nature of the interaction.

Groups revisited

Individuals sometimes form groups that have a permanence and uniformity of interaction with other groups that qualifies them to be thought of as individuals. This happens at various levels of evolution. We ourselves are such groups of cells. And humans participate in various social corpora—teams, states, ideological groups—that interact with others.

How such super-individuals are formed and hold together (or do not) is a central issue of both biology (R. Alexander 1979, 1987; Buss 1987; Maynard Smith and Szathmary 1995; Frank 1998, 2003) and social science. There is no one answer, but answers may involve both elements of cooperation and elements of spite. One important factor is the punishment of individuals who act against the interests of the group. A large experimental literature documents the willingness of many individuals to pay to punish "free riders" in public goods–provision games, and shows that such punishment is able to stabilize high levels of cooperation.[12] This is also an important finding of Ostrom's (1990) field studies of the self-organized government of commons. Costly punishment is, from an evolutionary point of view, a form of spite—although it is not called by that name in the literature.

The name may strike the reader as overly harsh in the case of the modest, graduated punishments found in Ostrom's field studies of successful cooperative collective management. But in tightly organized super-individuals, punishment can be draconian. Totalitarian regimes or ideologies classify those who violate social norms as traitors or heretics. They may be stoned to death. They have been burned at the

---

[11] A problem that one-population signaling models face in this regard arises when an individual interacts with another who sends the same signal. See Skyrms 2004.

[12] For instance, see Ostrom et al. 1992; and Fehr and Gachter 2000, 2002. Costly punishment is already implicit in the behavior of receivers in ultimatum-game experiments from Güth et al. 1982 to Henrich et al. 2004.

stake. The righteous people carrying out such acts no doubt believe that they are engaged in "altruistic punishment." We need also to think about the dark side of punishment.

When groups that can operate more or less like super-individuals have been formed, the interactions of the super-individuals themselves are also liable to the effects of positive and negative correlation described above. They can cooperate to produce a common good, or not. Their interactions may exemplify spite—not only in behavior, like the bacteria, but also in the full psychological sense of the word.

Repeated interactions, alliances, local interaction on a geographical landscape, signals, tags, and network formation all play a role. Division of labor is facilitated by trade networks, and trade may promote both the good and the bad sides of negative correlation.

The negative correlation conducive to spite in small populations may take on a larger significance when we consider interactions between groups. A local population of six interacting nations is perhaps more plausible than a local population of six interacting mice.

Evolution and the social contract

An evolutionary theory of the social contract stands in some contrast with social-contract theory as practiced in the contemporary philosophical treatment of John Harsanyi and John Rawls. They assume that everyone is, in some sense, rational. And they assume that in the relevant choice situation—behind a veil of ignorance—the relevant choosers are all basically the same. They all have the same rational-choice rule,[13] they all have the same basic values, and therefore they all make the same choice.[14] Correlation of types plays no part because it is assumed that there is only one relevant type.

Evolutionary game theory brings different types of individuals into the picture from the beginning. Evolutionary game theory is full of contingency. There are typically many equilibria; there are many possible alternative social contracts. The population might never get to equilibrium but rather cycle or describe a chaotic orbit. Mutation, invention,

---

[13] But theorists disagree about the nature of rational choice. Rawls minimizes the maximum loss; Harsanyi maximizes the expected payoff.

[14] The theorist tells you what that choice will be.

experimentation, and external environmental shocks add another layer of contingency.

Evolutionary game theory has some affinity with rational-choice theory in the absence of correlation.[15] This vanishes when interactions are correlated. But correlation, positive and negative, is the heart of the social contract. Correlation gets it started. Correlation lets it grow and develop more complex forms. Social institutions and networks evolve to enable and maintain correlation. Correlation explains much of what is admirable and what is despicable in existing social contracts—what we would like to keep and what we would like to change. A better understanding of the dynamics of correlation should be a central concern for Darwinian social philosophy.

Acknowledgments

I would like to thank my discussants at the Tanner symposium, Eleanor Ostrom, Michael Smith, and Peyton Young, for valuable comments on the lecture. I would also like to thank Jeffrey Barrett, Louis Narens, Don Saari, Rory Smead, Elliott Wagner, Jim Woodward, and Kevin Zollman for comments on earlier drafts, which much improved the lecture.

References

Alexander, J. M. (2000) "Evolutionary Explanations of Distributive Justice." *Philosophy of Science* 67: 490–516.

Alexander, J. M. (2007) *The Structural Evolution of Morality.* Cambridge: Cambridge University Press.

Alexander, J. M. and B. Skyrms (1999) "Bargaining with Neighbors: Is Justice Contagious?" *Journal of Philosophy* 96: 588–98.

Alexander, R. D. (1979) *Darwinism and Human Affairs.* Seattle: University of Washington Press.

Alexander, R. D. (1987) *The Biology of Moral Systems.* New York: de Gruyter.

Axelrod, R. (1981) "The Emergence of Cooperation among Egoists." *American Political Science Review* 75: 306–18.

Axelrod, R. (1984) *The Evolution of Cooperation.* New York: Basic Books.

---

[15] In large populations, expected fitness can then be calculated using population proportions in place of the subjective probabilities of rational choice theory.

Axelrod, R. and W. D. Hamilton (1981) "The Evolution of Cooperation." *Science* 211: 1390–6.

Axtell, R., J. M. Epstein, and H. P. Young (2006) "The Emergence of Classes in a Multi-agent Bargaining Model." In *Generative Social Science: Studies in Agent-Based Computational Modeling*, 177–95. Princeton: Princeton University Press.

Bergstrom, T. (2002) "Evolution of Social Behavior: Individual and Group Selection Models." *Journal of Economic Perspectives* 16: 231–38.

Bergstrom, T. and O. Stark (1993) "How Altruism Can Prevail in an Evolutionary Environment." *American Economic Review* 83: 149–55.

Binmore, K. (1994) *Game Theory and the Social Contract I: Playing Fair.* Cambridge, MA: MIT Press.

Binmore, K. (1998) *Game Theory and the Social Contract II: Just Playing.* Cambridge, MA: MIT Press.

Binmore, K. (2005) *Natural Justice.* Oxford: Oxford University Press.

Binmore, K., L. Samuelson, and H. P. Young (2003) "Equilibrium Selection in Bargaining Models." *Games and Economic Behavior* 45: 296–328.

Björnerstedt, J. and J. W. Weibull (1996) "Nash Equilibrium and Evolution by Imitation." In *The Rational Foundations of Economic Behavior*, ed. K. J. Arrow et al. New York: St. Martin's Press.

Bonacich, P. and T. Liggett (2003) "Asymptotics of a Matrix-Valued Markov Chain Arising from Sociology." *Stochastic Processes and Their Applications* 104: 155–71.

Braithwaite, R. B. (1955) *The Theory of Games as a Tool for the Moral Philosopher.* Cambridge: Cambridge University Press.

Burlando, R. M. and F. Guala (2005) "Heterogeneous Agents in Public Goods Experiments." *Experimental Economics* 8: 35–54.

Buss, L. W. (1987) *The Evolution of Individuality.* Princeton: Princeton University Press.

Dewey, J. (1910) *The Influence of Darwin on Philosophy, and Other Essays in Contemporary Thought.* New York: Henry Holt.

Dieckmann, T. (1999) "The Evolution of Conventions with Mobile Players." *Journal of Economic Behavior and Organization* 38: 93–111.

Durrett, R. and S. Levin (1994) "The Importance of Being Discrete (and Spatial)." *Theoretical Population Biology* 46: 363–94.

Ellison, G. (1993) "Learning, Local Interaction, and Coordination." *Econometrica* 61: 1047–71.

Ellison, G. (2000) "Basins of Attraction, Long-Run Stochastic Stability, and the Speed of Step-by-Step Evolution." *Review of Economic Studies* 67: 17–45.

Ely, J. (2002) "Local Conventions." *Advances in Theoretical Economics* 2, no. 1.

Epstein, J. M. (2006) *Generative Social Science: Studies in Agent-Based Computational Modeling.* Princeton: Princeton University Press.

Eshel, I. and L. L. Cavalli-Sforza (1982) "Assortment of Encounters and the Evolution of Cooperativeness." *Proceedings of the National Academy of Sciences of the USA* 79: 331–5.

Eshel, I., E. Sansone, and A. Shaked (1999) "The Emergence of Kinship Behavior in Structured Populations of Unrelated Individuals." *International Journal of Game Theory* 28: 447–63.

Fehr, E. and S. Gachter (2000) "Cooperation and Punishment in Public Goods Experiments." *American Economic Review* 90: 980–94.

Fehr, E. and S. Gachter (2002) "Altruistic Punishment in Humans." *Nature* 415: 137–40.

Feldman, M. and E. Thomas (1987) "Behavior-Dependent Contexts for Repeated Plays in the Prisoner's Dilemma II: Dynamical Aspects of the Evolution of Cooperation." *Journal of Theoretical Biology* 128: 297–315.

Fischbacher, U. and S. Gächter (2006) "Heterogeneous Social Preferences and the Dynamics of Free-Riding in Public Goods." Working paper, University of Zurich.

Frank, S. A. (1995) "George Price's Contributions to Evolutionary Genetics." *Journal of Theoretical Biology* 175: 373–88.

Frank, S. A. (1998) *Foundations of Social Evolution.* Princeton: Princeton University Press.

Frank, S. A. (2003) "Perspective: Repression of Competition and the Evolution of Cooperation." *Evolution* 57: 693–705.

Fudenberg, D. and D. Levine (1998) *A Theory of Learning in Games.* Cambridge, MA: MIT Press.

Gardner, A. and S. A. West (2004) "Spite and the Scale of Competition." *Journal of Evolutionary Biology* 17: 1195–1203.

Gauthier, D. (1985) "Bargaining and Justice." *Social Philosophy and Policy* 2: 29–47.

Gauthier, D. (1986) *Morals by Agreement.* Oxford: Oxford University Press.

Gibbard, A. (1990) *Wise Choices, Apt Feelings: A Theory of Normative Judgement.* Cambridge, MA: Harvard University Press.

Grafen, A. (1984) "Natural Selection, Kin Selection, and Group Selection." In *Behavioral Ecology: An Evolutionary Approach*, ed. J. R. Krebs and N. B. Davies, 62–84. Sunderland, Mass.: Sinauer.

Grafen, A. (1985) "A Geometric View of Relatedness." In *Oxford Surveys in Evolutionary Biology*, ed. R. Dawkins and M. Ridley, 2: 28–89. Oxford: Oxford University Press.

Greif, A. (1989) "Reputations and Coalitions in Medieval Trade." *Journal of Economic History* 49: 857–82.

Greif, A. (2006) *Institutions and the Path to the Modern Economy: Lessons from Medieval Trade.* Cambridge: Cambridge University Press.

Güth, W., R. Schmittberger, and B. Schwartze (1982) "An Experimental Analysis of Ultimatum Bargaining." *Journal of Economic Behavior and Organization* 3: 367–88.

Hamilton, W. D. (1963) "The Evolution of Altruistic Behavior." *American Naturalist* 97: 354–6.

Hamilton, W. D. (1964) "The Genetical Evolution of Social Behavior I and II." *Journal of Theoretical Biology* 7: 1–52.

Hamilton, W. D. (1971) "Selection of Selfish and Altruistic Behavior in Some Extreme Models." In *Man and Beast*, ed. J. F. Eisenberg and W. S. Dillon, 59–91. Washington, D.C.: Smithsonian Institution Press.

Hamilton, W. D. (1995) *Narrow Roads of Gene Land.* Vol. 1, *Evolution of Social Behavior.* New York: W. H. Freeman.

Hampton, J. (1996) *Hobbes and the Social Contract Tradition.* Cambridge: Cambridge University Press.

Harms, W. (2001) "Cooperative Boundary Populations: The Evolution of Cooperation on Mortality Risk Gradients." *Journal of Theoretical Biology* 213: 299–313.

Harms, W. (2004) *Information and Meaning in Evolutionary Processes.* New York: Cambridge University Press.

Harms, W. and B. Skyrms (2007) "Evolution of Moral Norms." In *Oxford Handbook in the Philosophy of Biology*, ed. Michael Ruse. Oxford: Oxford University Press.

Harsanyi, J. (2007) *Essays on Ethics, Social Behaviour, and Scientific Explanation.* Dordrecht: Reidel.

Hegselmann, R. (1996) "Social Dilemmas in Lineland and Flatland." In *Frontiers of Social Dilemmas Research*, ed. W. B. G. Liebrand and D. Messick, 337–62. Berlin: Springer.

Henrich, J., R. Boyd, S. Bowles, C. Camerer, E. Fehr, and H. Gintis (2004) *Foundations of Human Sociality: Economic Experiments and Ethnographic Evidence from Fifteen Small-Scale Societies.* New York: Oxford University Press.

Hofbauer, J. and K. Sigmund (1998) *Evolutionary Games and Population Dynamics.* Cambridge: Cambridge University Press.

Iwasa, Y., M. Nakamaru, and S. A. Levin (1998) "Allelopathy of Bacteria in a Lattice Population: Competition between Colicin-Sensitive and Colicin-Producing Strains." *Evolutionary Ecology* 12: 785–802.

Johnstone, R. A. and R. Bshary (2004) "Evolution of Spite through Indirect Reciprocity." *Proceedings of the Royal Society of London B* 271: 1917–22.

Kandori, M. (1992) "Social Norms and Community Enforcement." *Review of Economic Studies* 59: 63–80.

Kavka, G. (1986) *Hobbesian Moral and Political Theory.* Princeton: Princeton University Press.

Kitcher, P. (1993) "The Evolution of Human Altruism." *Journal of Philosophy* 10: 497–516.

Liggett, T. M. and S. W. W. Rolles (2004) "An Infinite Stochastic Model of Social Network Formation." *Stochastic Processes and Their Applications* 113: 65–80.

Luce, R. D. and H. Raiffa (1957) *Games and Decisions.* New York: Wiley.

Maynard Smith, J. (1964) "Group Selection and Kin Selection." *Nature* 201: 1145–7.

Maynard Smith, J. (1982) *Evolution and the Theory of Games.* Cambridge: Cambridge University Press.

Maynard Smith, J. and E. Szathmary (1995) *The Major Transitions in Evolution.* Oxford: Oxford University Press.

Milgrom, P., D. North, and B. Weingast (1990) "The Role of Institutions in the Revival of Trade: The Law Merchant, Private Judges, and the Champagne Fairs." *Economics and Politics* 2: 1–23.

Nowak, M. A. and R. M. May (1992) "Evolutionary Games and Spatial Chaos." *Nature* 359: 826–9.

Nowak, M. A. and K. Sigmund (1998) "Evolution of Indirect Reciprocity by Image Scoring." *Nature* 393: 573–7.

Oechssler, J. (1997) "Decentralization and the Coordination Problem." *Journal of Economic Behavior and Organization* 32: 119–35.

Ostrom, E. (1990) *Governing the Commons.* Cambridge: Cambridge University Press.

Ostrom, E., J. Walker, and R. Gardner (1992) "Covenants with and without a Sword: Self-Governance Is Possible." *American Political Science Review* 86: 404–17.

Pacheco, J. M., A. Traulsen, and M. A. Nowak (2006) "Active Linking in Evolutionary Games." *Journal of Theoretical Biology* 243: 437–43.

Page, T., L. Putterman, and B. Unel (2005) "Voluntary Association in Public Good Experiments: Reciprocity, Mimicry, and Efficiency." *Economic Journal* 115: 1032–53.

Pemantle, R. and B. Skyrms (2004a) "Network Formation by Reinforcement Learning: The Long and the Medium Run." *Mathematical Social Sciences* 48: 315–27.

Pemantle, R. and B. Skyrms (2004b) "Time to Absorption in Discounted Reinforcement Models." *Stochastic Processes and Their Applications* 109: 1–12.

Pollock, G. B. 1989. "Evolutionary Stability in a Viscous Lattice." *Social Networks* 11: 175–212.

Price, G. R. (1970) "Selection and Covariance." *Nature* 227: 520–1.

Ratnieks, F. and K. Visscher (1989) "Worker Policing in the Honeybee." *Nature* 342: 796–7.

Rawls, J. (1957) "Justice as Fairness." *Journal of Philosophy* 54: 653–62.

Rawls, J. (1971) *A Theory of Justice.* Cambridge, MA: Harvard University Press.

Robson, A. J. (1990) "Efficiency in Evolutionary Games: Darwin, Nash, and the Secret Handshake." *Journal of Theoretical Biology* 144: 379–96.

Samuelson, L. (1997) *Evolutionary Games and Equilibrium Selection.* Cambridge, MA: MIT Press.

Santos, F. C., J. M. Pacheco, and T. Lenaerts (2006) "Cooperation Prevails When Individuals Adjust Their Social Ties." *PLoS Computational Biology* 2, 10: 1–6.

Scanlon, T. (1998) *What We Owe to Each Other.* Cambridge, MA: Harvard University Press.

Schelling, T. (1960) *The Strategy of Conflict.* Cambridge, MA: Harvard University Press.

Schlag, K. H. (1998) "Why Imitate and If So, How? A Boundedly Rational Approach to Multi-armed Bandits." *Journal of Economic Theory* 78: 130–56.

Skyrms, B. (1996) *Evolution of the Social Contract.* Cambridge: Cambridge University Press.

Skyrms, B. (2001) "The Stag Hunt." *Proceedings and Addresses of the American Philosophical Association* 75: 31–41.

Skyrms, B. (2004) *The Stag Hunt and the Evolution of Social Structure.* Cambridge: Cambridge University Press.

Skyrms, B. (2007) "Dynamic Networks and the Stag Hunt: Some Robustness Considerations." *Biological Theory* 2, 1: 1–3.

Skyrms, B. and R. Pemantle (2000) "A Dynamic Model of Social Network Formation." *Proceedings of the National Academy of Sciences of the USA* 97: 9340–6.

Skyrms, B. and R. Pemantle (2004) "Learning to Network." In *Probability in Science*, ed. E. Eells and J. Fetzer. Chicago, IL: Open Court.

Sober, E. and D. S. Wilson (1998) *Unto Others: The Evolution and Psychology of Unselfish Behavior.* Cambridge, MA: Harvard University Press.

Sugden, R. (1986) *The Economics of Rights, Co-operation, and Welfare.* Oxford: Basil Blackwell.

Trivers, R. (1971) "The Evolution of Reciprocal Altruism." *Quarterly Review of Biology* 46: 35–57.

Vanderschraaf, P. (1998) "The Informal Game Theory in Hume's Account of Convention." *Economics and Philosophy* 14: 215–47.

Vanderschraaf, P. (2006) "War or Peace: A Dynamical Analysis of Anarchy." *Economics and Philosophy* 22: 243–79.

Vanderschraaf, P. and J. M. Alexander (2005) "Follow the Leader: Local Interaction with Influence Neighborhoods." *Philosophy of Science* 72: 86–113.

Wahl, L. M. (2002) "Evolving the Division of Labor: Generalists, Specialists, and Task Allocation." *Journal of Theoretical Biology* 219: 371–88.

Weibull, J. (1995) *Evolutionary Game Theory.* Cambridge, MA: MIT Press.

Wright, S. (1921) "Systems of Mating III: Assortative Mating Based on Somatic Resemblance." *Genetics* 6: 144–61.

Wright, S. (1945) "Tempo and Mode in Evolution: A Critical Review." *Ecology* 26: 415–19.

Young, H. P. (1993a) "An Evolutionary Model of Bargaining." *Journal of Economic Theory* 59: 145–68.

Young, H. P. (1993b) "The Evolution of Conventions." *Econometrica* 61: 57–84.

Young, H. P. (1998) *Individual Strategy and Social Structure.* Princeton: Princeton University Press.

Zollman, K. (2005) "Talking to Neighbors: The Evolution of Regional Meaning." *Philosophy of Science* 72: 69–85.

PART II

Importance of Dynamics

Introduction

The chapters in this section argue for the importance of dynamic analysis. Other themes, of course, get mixed in. The first paper, "Trust, Risk, and the Social Contract" shows that consideration of network dynamics can account for trust without postulating any innate disposition to trust. This paper won a prize, perhaps for brevity. It provides a preview of coming attractions in Part III. If you like it, you may want to read more.

There is more that one sort of dynamics, and results may be sensitive to the dynamic setting. The previous paper showed this with respect to the Stag Hunt. "Bargaining with Neighbors: Is Justice Contagious?", written with Jason McKenzie Alexander, shows that the prospects for the emergence of the equal split in symmetric situations are much better when bargaining with neighbors than when bargaining with strangers. Local interaction makes all the difference, with clumps of egalitarians forming and then spreading to take over the population.

The phenomenon was discovered by simulation, but then Jason was able to give an analytic treatment by looking at interactions at the edges of clumps. If this should whet your interest in the importance of interactions with neighbors, I recommend Jason's book *The Structural Evolution of Morality*.

"Stability and Explanatory Significance of Some Simple Evolutionary Models" introduces the notion of the structural stability of a dynamic model. If a model is structurally unstable, then an arbitrary small perturbation in the dynamics can lead to large changes in the global behavior of the model. (That is a non-technical description. It is a precise notion.) The explanatory significance of a structurally unstable model may be rather suspect. An equilibrium may be dynamically stable in a model, but if the model is structurally unstable then with a slight change in the dynamics it may not be an equilibrium at all. I discuss dynamic stability, structural stability, and robustness to some larger perturbations in the dynamics, for three simple evolutionary models.

"Dynamics of Conformist Bias" explores perturbing evolutionary (replicator) dynamics by adding a little conformist bias. All other things being equal, a behavior that is more widely followed that the alternatives is more likely to gain population share. Conformist bias is an undeniable factor in cultural evolution. And in situations where the straight replicator model is structurally unstable, a very small amount of conformist bias may make a very big change. I find general statements about the benefits of conformist bias, however, premature. Conformist bias may have group benefits, but it may also have group detriments, depending on the kind of interaction that is evolving.

"Chaos and the Explanatory Significance of Equilibrium: Strange Attractors in Evolutionary Game Dynamics" makes the case for taking the dynamics seriously in the most dramatic way possible. Equilibrium analysis is not very good if one never gets to equilibrium. This paper exhibits a dynamic model that (1) never reaches an equilibrium, and (2) isn't even predictible. One might suspect that chaos in game dynamics is a rare, rigged-up phenomenon, but see Cowan (1992), Sato et al. (2002), and Galla and Farmer (2013). One might suspect that it does not occur in any games of real interest, but see Wagner (2012) for chaos in a signaling game.

"Evolutionary Dynamics of Collective Action in N-person Stag Hunt Dilemmas," written with Jorge Pacheco, Francisco Santos, and Max Souza, moves from two person games to N-person public goods provision games with a threshold. Public goods provision games had been usually presented as N-person Prisoner's Dilemmas in which there is only one equilibrium, the one in which no one cooperates. The idea of a threshold, where a certain number of the group needs to cooperate to produce any benefit, is quite natural in many situations. This can turn the N-person game from a version of the Prisoner's Dilemma into a version of the Stag Hunt. There are now multiple equilibria. Both infinite and finite population dynamics are studied and compared.

"Learning to Take Turns: The Basics," written with Peter Vanderschraaf, is a somewhat abridged and simplified version of our longer paper "Learning to Take Turns." We introduce a game dynamics called Markov Fictitious Play. This incorporates a simple kind of pattern learning. Players make a Bayesian inference about transitions, and then best respond to the resulting beliefs. In situations in which stable Nash equilibria may be inequitable, "taking turns" equilibria afford an equitable arrangement. Our pattern-recognizing players are capable of learning to take turns in these situations. They are not guaranteed to do so, but they may do so. The dynamics corresponds to real situations in which individuals may find it difficult to learn to take turns, but may nevertheless do so. For theorems and simulations estimating basins of attraction of taking-turns-equilibria, see the longer paper and Peter's book *Learning and Coordination* (1991).

"Evolutionary Considerations in the Framing of Social Norms" was written with Kevin Zollman for a symposium on Cristina Bicchieri's book, *The Grammar of Society*. Our paper puts together two ideas. The first is that norms evolve for classes of games, not individual games. These classes may be nested, so that a particular interaction falls within several classes. The second is that framing can be modeled as sending a signal that a situation falls within a certain class. The signal tends to trigger the norm that was evolved for that class. This account can explain a variety of puzzling experimental phenomena in a uniform way.

References

Alexander, J. M. (2007) *The Structural Evolution of Morality*. Cambridge: Cambridge University Press.

Bicchieri, C. (2005) *The Grammar of Society: The Nature and Dynamics of Social Norms*. Cambridge: Cambridge University Press.

Cowan, S. G. (1992) "Dynamical Systems arising from Game Theory." Ph.D. thesis, University of California at Berkeley.

Galla, T. and J. D. Farmer (2013) "Complex Dynamics in Learning Complicated Games." *PNAS* 110: 1232–6.

Sato, Y., E. Akiyama, and J. D. Farmer (2002) "Chaos in Learning a Simple Two-Person Game." *PNAS* 99: 4748–51.

Vanderschraaf, P. (1991) *Learning and Coordination*. New York: Routledge.

Wagner, E. (2012) "Deterministic Chaos and the Evolution of Meaning." *British Journal for the Philosophy of Science* 63: 547–75.

2

Trust, Risk, and the Social Contract

Two neighbors may agree to drain a meadow, which they possess in common; because 'tis easy for them to know each other's mind, and each may perceive that the immediate consequence of failing in his part is the abandoning of the whole project. But 'tis difficult, and indeed impossible, that a thousand persons shou'd agree in any such action.

David Hume *A Treatise of Human Nature*,
Bk. III, Pt.II, Sec. VII.

Social contracts, great and small, depend on trust. Hume's two neighbors are able to sustain an implicit contract even though the failure of one to perform his part causes the cooperative enterprise to fail. If we view Hume's two neighbors through the lens of game theory, the simplest representation is that they are playing a two-person non-zero sum game with the structure of a Stag Hunt. (The name comes from a story in Rousseau with a similar moral.) There are two equilibria: *both cooperate*; *neither cooperates*. It is not an equilibrium for one to cooperate and the other not, because in such a case each would have an incentive to switch. The equilibrium where both cooperate is the one in which they are both better off—it is said to be *payoff dominant*—but each runs the risk that the other may not do his part. The equilibrium where neither cooperates is one in which neither player runs a risk, since the outcome is the same no matter what the other player does. (In Rousseau's story a successful Stag Hunt requires cooperation, while Hare Hunting is a solitary occupation.) Mutual benefit is pitted against the risk that the other (or others) may not honor the implicit contract. This is the prototypical problem of the social contract.

How bad is the problem? That depends on two things: first—as Hume points out—the beliefs about what the other player does; second, the magnitudes of the benefit of cooperation and of the risk of being abandoned by one's potential partner.

Suppose that the effort in doing one's part is E, the benefit to each of a drained meadow is B, and the value of the status quo is D. Then the meadow draining game has payoffs (of form row's payoff, column's payoff):

	Work to drain	Don't work
Work to drain	B–E, B–E	D–E, D
Don't work	D, D–E	D, D

If the project is worth the effort, B–E > D, this has the structure of a Stag Hunt game. If I am sure that I can trust my partner to cooperate, I will too.

But suppose that my partner is as likely to work as not. Then my expectation of working is $(1/2)(B–E) + (1/2)(D–E)$ and my expectation for not working is D. Here I am indifferent between working or not if B–D = 2E. A little better benefit or a little less required effort tips the balance in favor of cooperation; a little less benefit or a little increase in the work load tips the balance the other way. In any case, the equilibrium that comes out ahead in this calculation—with equal probabilities for the other cooperating or not—is called the *risk dominant* equilibrium. In easy cases, the risk dominant equilibrium coincides with the payoff dominant equilibrium; in hard cases risk dominance and payoff dominance pull in the opposite direction. Let us direct our attention to hard cases.

For example, suppose that B = 7, E = 3, and D = 3. Then we have a Stag Hunt game in which draining the meadow is the payoff-dominant equilibrium and no one working is the risk dominant equilibrium:

	Work to drain (Stag)	Don't work (Hare)
Work to drain (Stag)	4,4	0,3
Don't work (Hare)	3,0	3,3

If people have a moderate degree of trust, probability > 0.75, cooperation ensues. But where does the trust come from? One can say that trust is based on prior experience, but that only pushes the question back. If there had not been prior trust, then there would not have been the kind of prior experience that supports trust, but rather the kind that supports distrust.

Perhaps, someone might say, we have evolved to be the kind of species with a predilection for cooperation—with some initial but defeasible predilection for trust in cooperative enterprises built into our nature. The same problem now emerges in even grander evolutionary terms. Should we expect evolutionary dynamics to respect payoff dominance when it conflicts with risk dominance?

The answer usually delivered up by contemporary evolutionary game theory is "No." In the long run, one should expect to see the risk-dominant equilibrium almost all the time. This is the central result of Kandori et al. (1993)[1] (See also Young 1998). The idea is that there is an underlying dynamics of differential reproduction perturbed by some very small probability of mutation. Sooner or later—perhaps a lot later—a lot of mutations move the (finite) population from the basin of attraction of one equilibrium to that of the other, and then the underlying differential reproduction quickly takes it to that equilibrium. Sooner or later, a lot of mutations take the population to the basin of attraction of the other equilibrium, and differential reproduction takes the population to the second equilibrium. If mutations are unlikely and the probability of mutations is independent across individuals, the probability of mutations taking you from the basin of attraction of the cooperative equilibrium to that of the non-cooperative one is much larger than the probability of mutations taking you in the opposite direction. Therefore the population spends most of its time not cooperating.

On the face of it, the reasoning seems remarkably robust. The underlying dynamics need not be differential reproduction; it could be anything with the same basin of attraction—anything that moves in the direction of greatest payoff. The stochastic shocks to the system might be

---

[1] But compare Robson and Vega-Redondo (1996). They point out that Kandori–Mailath–Rob assume a special type of matching (round-robin matching) to get their results, and show how the correlations generated by random matching in a finite population can lead to the payoff dominant equilibrium in the Stag Hunt game. Later in this chapter we consider correlations in matching that arise not by chance, but by the choices of the agents involved.

interpreted as experimentation, or as some kind of exogenous noise with a similar probability structure (see Foster and Young 1990). The only problem seems to be that the expected waiting time for all those mutations, experiments or whatever to happen all at once may be astronomical. It is almost like a theory of evolution driven by the probability of miracles.

Ellison (1993) provides an answer from perhaps an unexpected quarter. He considers a local interaction model where individuals are located on a circle, and interact with their immediate neighbors. It is still true that with a small probability of mutations, the population spends almost all its time not cooperating, but here the expected waiting time is short. If mutation delivers two contiguous defectors in a population of cooperators, the defectors will rapidly spread and take over the population. In a later paper Ellison (2000) shows that defecting also takes over quickly—though not quite so quickly—in a two-dimensional local interaction model where individuals play with their neighbors to the North, East, South, and West.[2] Young (1998) gives further support to the selection of risk-dominant equilibria in games played on local interaction structures. We do not yet have a good model for the emergence of sufficient trust to allow the selection of the mutually beneficial equilibrium over the risk-dominant equilibrium in a coordination game.

Why not? It is because the foregoing analysis holds the interaction structure *fixed*. We should consider the possibility that individuals learn *with whom* to interact as well as *how* to act. Interaction structures will then be dynamic entities, and strategy and structure will co-evolve. Such a model is proposed in Skyrms and Pemantle (2000, Chapter 11 in this volume), and pursued in Pemantle and Skyrms (2004a, Chapter 12, 2004b, Chapter 13), Skyrms (2004) and Skyrms and Pemantle (2004). In this model a small group of individuals start interacting at random. Interactions are modeled as games, and agents' types determine their strategies in the games. The interaction structure evolves by reinforcement learning, with the magnitude of reinforcements being the payoffs

---

[2] However, in contrast to the original Kandori–Mailath–Rob and Foster–Young models, the underlying deterministic dynamics here *does* matter. Ellison uses a best-response dynamics. If you switch to an imitate-the-best dynamics, for example, you get quite different results (see Skyrms 2004: ch. 3).

from interactions. If you have a good payoff from an interaction with someone, you are more likely to interact with them again.

Consider our Stag Hunt game in this context, where Stag hunting (cooperating) is payoff dominant and Hare hunting (not-cooperating) is risk-dominant. Using a standard model of reinforcement learning (Roth and Erev 1995), Stag hunters rapidly learn to interact with one another. In our prototypical Stag Hunt, Stag hunters then get a payoff of 4 while Hare hunters get a payoff of 3. Risk-dominance now has no teeth, because a little learning has taken the risk out of Stag hunting.

Since Stag hunters are now doing better than Hare hunters, Hare hunters may eventually notice this and imitate them—or in a biological context Stag hunters can out-reproduce Hare hunters. In either case Stag hunting takes over the population. All that is required is that the interaction structure is sufficiently fluid so that Stag hunters can find each other quickly.

The choice of reinforcement learning as the dynamics of interaction requires little of the individuals involved. They don't need to think strategically about the situation at all, and they don't need to observe others' actions or payoffs. More sophisticated and knowledgeable Stag hunters who are free to associate as they please might find each other *right away*. The model of reinforcement learning is therefore something of a worst case analysis.

The foregoing "Learning-to-Network" model was designed for a small group in which agents can identify one another and keep track of reinforcements associated with an individual. This may not be plausible in large populations—but large populations may be made up of smaller subgroups. Suppose that a large population consists of a number of small groups—demes—within which interactions take place and interaction structure is adjusted by learning. Strategies may be adjusted by imitation or may also be adjusted by reinforcement. The learning rates may vary from deme to deme. If learning with whom to interact is fast relative to strategy revision, Stag hunting predominates. In demes where interaction structure is rigid and Stag hunters are repeatedly let down by Hare hunters but strategy revision is fast, people learn to hunt Hare.

Now suppose that people can move from one locality to another— perhaps at some small cost, know a little about the norms of other demes, and can do just a little bit of strategic thinking. Stag hunters stuck in a noncooperative deme can now move to Stag-hunting demes. Stag hunters associate with one another on a larger scale.[3] This effect alone could guarantee the eventual success of Stag hunting even if within-deme dynamics were only driven by mutation and random interaction (see Oechssler 1999; Ely 2002; Dieckmann 1999). But one would have to wait until good luck got the process started.[4] Learning-to-Network within demes jump-starts cooperation, which can then spread to a larger population through mobility between demes.

What did we need to explain the possibility of cooperation in the Stag Hunt game? We did not need to assume that evolution had to somehow built in a disposition for trust. It is enough that agents come equipped with a modest capacity for learning.

References

Alexander, J. M. (2003) "Random Boolean Networks and Evolutionary Game Theory." *Philosophy of Science* 70: 1289–304.

Dieckmann, T. (1999) "The Evolution of Conventions with Mobile Players." *Journal of Economic Behavior and Organization* 38: 93–111.

Ellison, G. (1993) "Learning, Local Interaction, and Coordination." *Econometrica* 61: 1047–71.

Ellison, G. (2000) "Basins of Attraction, Long-run Stochastic Stability, and the Speed of Step-by-step Evolution." *Review of Economic Studies* 67: 17–45.

Ely, J. (2002) "Local Conventions." *Advances in Theoretical Economics* 2.

Foster, D. and H. P. Young (1990) "Stochastic Evolutionary Game Dynamics." *Theoretical Population Biology* 38: 219–22.

Kandori, M., Mailath, G., and R. Rob (1993) "Learning, Mutation and Long Run Equilibria in Games." *Econometrica* 61: 29–56.

Oechssler, J. (1999) "Competition among Conventions." *Mathematical and Computational Organization Theory* 5: 31–44.

Pemantle, R. and B. Skyrms (2004a) "Network Formation by Reinforcement Learning: the Long and the Medium Run." *Mathematical Social Sciences* 48: 315–27.

Pemantle, R. and B. Skyrms (2004b) "Time to Absorption in Discounted Reinforcement Models." *Stochastic Processes and Their Applications* 109: 1–12.

---

[3] Hare hunters have no reason to move, and may be left by themselves in poor Hare hunting villages or gradually be converted to Stag hunting, depending on how far they can see and how they revise strategies (see Skyrms 2004: ch. 7).

[4] If set up just right, these models can be very fast, but alternative versions can be very slow.

Robson, A. J. and F. Vega-Redondo (1996) "Efficient Equilibrium Selection in Evolutionary Games with Random Matching." *Journal of Economic Theory* 70: 65–92.

Roth, A. and I. Erev (1995) "Learning in Extensive Form Games: Experimental Data and Simple Dynamic Models in the Intermediate Term." *Games and Economic Behavior* 8: 164–212.

Rousseau, J. (1984) *A Discourse on Inequality*. Trans. M. Cranston. New York: Penguin Books.

Skyrms, B. (2004) *The Stag Hunt and the Evolution of Social Structure*. New York: Cambridge University Press.

Skyrms, B. and R. Pemantle (2000) "A Dynamic Model of Social Network Formation." *Proceedings of the National Academy of Sciences of the USA* 97: 9340–6.

Skyrms, B. and R. Pemantle (2004) "Learning to Network." In *Probability in Science* ed. E. Eells and J. Fetzer. Chicago, IL: Open Court Publishing.

Vanderschraaf, P. and J. M. Alexander (2005) "Follow the Leader: Local Interactions with Influence Neighborhoods." *Philosophy of Science* 72: 86–113.

Young, H. P. (1993) "The Evolution of Conventions." *Econometrica* 61: 57–84.

Young, H. P. (1998) *Individual Strategy and Social Structure*. Princeton, NJ: Princeton University Press.

3

Bargaining With Neighbors: Is Justice Contagious?

with Jason Alexander

What is justice? The question is harder to answer in some cases than in others. We focus on the easiest case of distributive justice. Two individuals are to decide how to distribute a windfall of a certain amount of money. Neither is especially entitled, or especially needy, or especially anything—their positions are entirely symmetric. Their utilities derived from the distribution may be taken, for all intents and purposes, simply as the amount of money received. If they cannot decide, the money remains undistributed and neither gets any. The essence of the situation is captured in the simplest version of a bargaining game devised by John Nash (1950). Each person decides on a bottom-line demand. If those demands do not jointly exceed the windfall, then each person gets his demand; if not, no one gets anything. This game is often simply called *divide-the-dollar*.

In the ideal simple case, the question of distributive justice can be decided by two principles:

- *Optimality*: a distribution is not just if, under an alternative distribution, all recipients would be better off.
- *Equity*: if the position of the recipients is symmetric, then the distribution should be symmetric. That is to say, it does not vary when we switch the recipients.

Since we stipulate that the position of the two individuals is symmetric, equity requires that the just distribution must give them the same amount of money. Optimality then rules out such unlikely schemes as giving each one dime and throwing the rest away—each must get half the money.

There is nothing new about our two principles. Equity is the simplest consequence of the theory of distributive justice in Aristotle's *Politics*. It is a consequence of Immanuel Kant's categorical imperative. Utilitarians tend to stress optimality, but are not completely insensitive to equity. Optimality and equity are the two most uncontroversial requirements in Nash's axiomatic treatment of bargaining. If you ask people to judge the just distribution, their answers show that optimality and equity are powerful operative principles (Yaari and Bar-Hillel 1981). So, although nothing much hangs on it, we shall feel free to use moral language and to call the equal split *fair division* in divide-the-dollar.

1. Rationality, Behavior, Evolution

Two rational agents play the divide-the-dollar game. Their rationality is common knowledge. What do they do? The answer that game theory gives us is that *any* combination of demands is compatible with these assumptions. For example, Jack may demand 90 percent thinking that Jill will only demand 10 percent on the assumption that Jill thinks that Jack will demand 90 percent and so forth, while Jill demands 75 percent thinking that Jack will demand 25 percent on the assumption that Jack thinks that Jill will demand 75 percent and so forth. *Any* pair of demands is *rationalizable*, in that it can be supported by a hierarchy of conjectures for each player, compatible with common knowledge of rationality. In the example given, these conjectures are quite mistaken.

Suppose we add the assumption that each agent somehow knows what the other will demand. Then any combination of demands that total the whole sum to be divided is still possible. For example, suppose that Jack demands 90 percent knowing that Jill will demand 10 percent and Jill demands 10 percent knowing that Jack will demand 90 percent. Then each player is maximizing payoff given the demand of the other. That is to say that this is a Nash equilibrium of divide-the-dollar. If the dollar is infinitely divisible, then there are an infinite number of such equilibria.

If experimental game theorists have people actually play divide-the-dollar, they *always* split equally (van Huyck et al. 1955; Nydegger and Owen 1974; Roth and Malouf 1979). This is not always true in more complicated bargaining experiments where there are salient asymmetries, but it is true in divide-the-dollar. Rational-choice theory has no explanation of this phenomenon. It appears that the experimental subjects are using norms of justice to select a particular Nash equilibrium of the game. But what account can we give for the existence of these norms?

Evolutionary game theory (reading 'evolution' as cultural evolution) promises an explanation, but the promise is only partially fulfilled. Demand-half is the only evolutionarily stable strategy in divide-the-dollar (Sugden 1986). It is the only strategy such that, if the whole population played that strategy, no small group of innovators, or "mutants," playing a different strategy could achieve an average payoff at least as great as the natives. If we could be sure that this unique evolutionarily stable strategy would always take over the population, the problem would be solved.

But we cannot be sure that this will happen. There are states of the population that are evolutionarily stable where some fraction of the population makes one demand and some fraction makes another. The state where half the population demands one-third and half the population demands two-thirds is such an evolutionarily stable polymorphism of the population. So is the state where two-thirds of the population demands 40 percent and one-third of the population demands 60 percent. We can think of these as pitfalls along the evolutionary road to justice.

How important are these polymorphisms? To what extent do they compromise the evolutionary explanation of the egalitarian norm? We cannot begin to answer these questions without explicitly modeling the evolutionary dynamics and investigating the size of their basins of attraction.

2. Bargaining with Strangers

The most widely studied dynamic evolutionary model is a model of interactions with strangers. Suppose that individuals are paired at random from a very large population to play the bargaining game. We assume that the probability of meeting a strategy can be taken as the proportion of the population that has that strategy. The population proportions evolve according to the replicator dynamics. The proportion of the population using a strategy in the next generation is the proportion playing that strategy in the current generation multiplied by a *fitness factor*. This fitness factor is just the ratio of the average payoff to this strategy to the average payoff in the whole population.[1] Strategies that do better than average grow; those which do worse than average shrink. This dynamic arose in biology as a model of asexual reproduction, but more to the point here, it also has a cultural evolutionary interpretation where strategies are imitated in proportion to their success (Björnerstedt and Weibull 1966; Schlag 1966).

The basins of attraction of these polymorphic pitfalls are not negligible. A realistic version of divide-the-dollar will have some finite number of strategies instead of the infinite number that we get from the idealization of infinite divisibility. For a finite number of strategies, the size of a basin of attraction of a population state makes straightforward sense. It can be estimated by computer simulations. We can consider coarse-grained or fine-grained versions of divide-the-dollar; we can divide a stack of quarters, or of dimes, or of pennies. Some results of simulations persist across a range of different granularities. Equal division always has the largest basin of attraction and it is always greater than the basins of attractions of all the polymorphic pitfalls combined. If you choose an initial population state at random, it is more probable than not that the replicator dynamics will converge to a state of fixation of demand-half. Simulation results range between 57 and 63 percent of the initial points going to fair division. The next largest basin of attraction is always that closest to the equal split: for example, the four–six polymorphism in the case of dividing a stack of ten dimes and the forty-nine–fifty-one polymorphism in the case of dividing a stack of one-hundred pennies. The rest of the polymorphic equilibria follow the general rule—the closer to fair division, the larger the basin of attraction.

For example, the results running the discrete replicator dynamics to convergence and repeating the process 100,000 times on the game of dividing 10 dimes are given in Table 3.1.

The projected evolutionary explanation seems to fall somewhat short. The best we might say on the basis of pure replicator dynamics is that

---

[1] This is the discrete time version of the replicator dynamics, which is most relevant in comparison to the alternative *bargaining-with-neighbors* dynamics considered here. There is also a continuous time version. As comprehensive references, see Hofbauer and Sigmund (1988), Weibull (1995), and Samuelson (1997).

Table 3.1 Convergence results for replicator dynamics—
100,000 trials

Fair Division	62,209
4–6 Polymorphism	27,469
3–7 Polymorphism	8,801
2–7 polymorphism	1,483
1–9 Polymorphism	38
0–10 Polymorphism	0

fixation of fair division is more likely than not, and that polymorphisms far from fair division are quite unlikely.

We can say something more if we inject a little bit of probability into the model. Suppose that every once and a while a member of the population just picks a strategy at random and tries it out—perhaps as an experiment, perhaps just as a mistake. Suppose we are at a polymorphic equilibrium—for instance, the four–six equilibrium in the problem of dividing ten dimes. If there is some fixed probability of an experiment (or mistake), and if experiments are independent, and if we wait long enough, there will be enough experiments of the right kind to kick the population out of the basin of attraction of the four–six polymorphism and into the basin of attraction of fair division and the evolutionary dynamics will carry fair division to fixation. Eventually, experiments or mistakes will kick the population out of the basin of attraction of fair division, but we should expect to wait much longer for this to happen. In the long run, the system will spend most of its time in the fair-division equilibrium. Peyton Young (1993a; see also Foster and Young 1990; Young 1993b) showed that, if we take the limit as the probability of someone experimenting gets smaller and smaller, the ratio of time spent in fair division approaches one. In his terminology, fair division is the *stochastically stable equilibrium* of this bargaining game.

This explanation gets us a probability arbitrarily close to one of finding a fair-division equilibrium if we are willing to wait an arbitrarily long time. But one may well be dissatisfied with an explanation that lives at infinity. (Putting the limiting analysis to one side, pick some plausible probability of experimentation or mistake and ask yourself how long you would expect it to take in a population of 10,000, for 1,334 demand-six types simultaneously to try out being demand-five types and thus kick the population out of the basin of attraction of the four–six polymorphism and into the basin of attraction of fair division.[2]) The evolutionary explanation still seems less than compelling.

3. Bargaining with Neighbors

The model of random encounters in an infinite population that motivates the replicator dynamics may not be the right model. Suppose interactions are with neighbors. Some investigations of cellular automaton models of Prisoner's Dilemma and a few other games show that interactions with neighbors may produce dynamical behavior quite different from that generated by interactions with strangers (Pollack 1989; Nowak and May 1992; Lindgren and Nordahl 1994; Anderlini and Ianni 1997). Bargaining games with neighbors have not, to the best of our knowledge, previously been studied.

Here, we investigate a population of 10,000 arranged on a one hundred by one hundred square lattice. As the neighbors of an individual in the interior of the lattice, we take the eight individuals to the N, NE, E, SE, S, SW, W, NW. This is called the Moore(8) neighborhood in the cellular automaton literature.[3] The dynamics is driven by imitation. Individuals imitate the most successful person in the neighborhood. A generation—an iteration of the discrete dynamics—has two stages. First, each individual plays the divide-ten-dimes game with each of her neighbors using her current strategy. Summing the payoffs gives her current success level. Then each player looks around her neighborhood and changes her current strategy by imitating her most successful neighbor, providing that her most successful neighbor is more successful than she is; otherwise, she does not switch strategies. (Ties are broken by a coin flip.)

In initial trials of this model, fair division *always* went to fixation. This cannot be a universal law, since you can design "rigged" configurations where a few demand-one-half players are, for example, placed in a population of demand-four and demand-six players with the latter so

---

[2] For discussion of expected waiting times, see Ellison (1993) and Axtell et al. (1999).
[3] We find that behavior is not much different if we use the von Neumann neighborhood: N, S, E, W, or a larger Moore neighborhood.

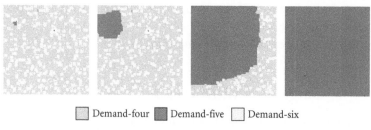

☐ Demand-four ■ Demand-five ☐ Demand-six

Figure 3.1 The steady advance of fair division

arranged that there is a demand-six type who is the most successful player in the neighborhood of every demand-one-half player. Start enough simulations at random starting points and sooner or later you will start at one of these.

We ran a large simulation starting repeatedly at randomly chosen starting points. Fair division went to fixation in more than 99.5 percent of the trials. The cases where it did not were all cases where the initial population of 10,000 contained fewer than seventeen demand-one-half players. Furthermore, convergence was remarkably quick. Mean time to fixation of fair division was about sixteen generations. This may be compared with a mean time to convergence[4] in discrete replicator dynamics of forty-six generations, and with the ultra-long-run character of stochastically stable equilibrium.

It is possible to exclude fair division from the possible initial strategies in the divide-ten-dimes game and start at random starting points that include the rest. If we do this, all strategies other than demand-four dimes and demand-six dimes are eliminated and the four–six polymorphic population falls into a "blinking" cycle of period two. If we then turn on a little bit of random experimentation or "mutation" allowing the possibility of demand-five, we find that as soon as a very small clump of demand-five players arises, it systematically grows until it takes over the whole population—as illustrated in Figure 3.1. *Justice is contagious.*[5]

---

[4] At .9999 level to keep things comparable.
[5] Ellison (1993) found such contagion effects in local interaction of players arranged on a circle and playing pure coordination games.

4. Robustness

The bargaining-with-neighbors model of the last section differs from the bargaining-with-strangers model in more than one way. Might the difference in behavior that we have just described be due to the imitate-the-most-successful dynamics rather than the neighbor effect? To answer this question, we ran simulations varying these factors independently.

We consider both fixed and random neighborhoods. The models using fixed neighborhoods use the Moore (8) neighborhood described above. In the alternative random-neighborhood model, in each generation a new set of "neighbors" is chosen at random from the population for each individual. That is to say, these are neighborhoods of *strangers*.

We investigated two alternative dynamics. One imitates the most successful neighbor as in our bargaining-with-neighbors model. The other tempers the all-or-nothing character of imitate-the-best. Under it, an individual imitates one of the strategies in its neighborhood that is more successful than it (if there are any) with relative probability proportional to their success in the neighborhood. This is a move in the direction of the replicator dynamics.

In Table 3.2, A and B are bargaining with neighbors, with imitate-the-best-neighbor and imitate-with-probability-proportional-to-success dynamics, respectively. The results are barely distinguishable. C and D are the random-neighborhood models corresponding to A and B, respectively. These results are much closer to those given for the replicator dynamics in Table 3.1. The dramatic difference in convergence to fair division between our two models is due to the structure of the interaction with neighbors.

Table 3.2 Convergence results for five series of 10,000 trials

	Bargaining with neighbors		Bargaining with strangers	
	A	B	C	D
0–10	0	0	0	0
1–9	0	0	0	0
2–8	0	0	54	57
3–7	0	0	550	556
4–6	26	26	2560	2418
fair	9972	9973	6833	6964

5. Analysis

Why is justice contagious? A strategy is contagious if an initial "patch" of that strategy will extend to larger and larger patches. The key to the contagion of a strategy is interaction along the edges of the patch, since in the interior the strategy can only imitate itself.[6]

Consider an edge with demand-five players on one side, and players playing the complementary strategies of one of the polymorphisms on the other. Since the second rank of demand-five players always meet their own kind, they each get a total payoff of forty from their eight neighbors. Players in the first rank will therefore imitate them unless a neighbor from the polymorphism gets a higher payoff. The low strategy in a polymorphic pair cannot get a higher payoff. So if demand-five is to be replaced at all, it must be by the high strategy of one of the polymorphic pairs.

In the four–six polymorphism—the polymorphism with the greatest basin of attraction in the replicator dynamics—this simply cannot happen, even in the most favorable circumstances. Suppose that we have someone playing demand-six in the first rank of the polymorphism, surrounded on his own side by compatible demand-four players to boost his payoff to the maximum possible.[7] Since he is in the first rank, he faces three incompatible demand-five neighbors. He has a total payoff of thirty while his demand-five neighbors have a total payoff of thirty-five. Demand-five begins an inexorable march forward as illustrated in Figure 3.2. (The pattern is assumed to extend in all directions for the computation of payoffs of players at the periphery of what is shown in the figure.)

If we choose a polymorphism that is more extreme, however, it is possible for the high strategy to replace some demand-five players for a while. Consider the one–nine polymorphism, with a front line demand-nine player backed by compatible demand-one neighbors. The demand-nine player gets a total payoff of forty-five—more than anyone else—and

---

[6] For this reason, "frontier advantage" is used to define an unbeatable strategy in Eshel et al. (1996).

[7] In situating the high strategy of the polymorphic pair in a sea of low-strategy players, we are creating the best-case scenario for the advancement of the polymorphism into the patch of demand-five players.

Initial		Iteration 1
5544		5554
5544		5554
5564	=>	5554
5544		5554
5544		5554

Figure 3.2 Fair division versus four–six polymorphism

Initial		Iteration 1		Iteration 2		Iteration 3
55111		55511		55551		55555
55111		55511		55559		55559
55111	=>	59991	=>	55999	=>	55599
55911		59991		55999		55599
55111		59991		55999		55599
55111		55511		55559		55559
55111		55511		55551		55555

Figure 3.3 Fair division versus one–nine polymorphism

thus is imitated by all his neighbors. This is shown in the first transition in Figure 3.3.

But the success of the demand-nine strategy is its own undoing. In a cluster of demand-nine strategies, it meets itself too often and does not do so well. In the second transition, demand-five has more than regained its lost territory, and in the third transition it has solidly advanced into one–nine territory.

Analysis of the interaction along an edge between demand-five and other polymorphisms is similar to one of the cases analyzed here.[8] Either the polymorphism cannot advance at all, or the advance creates the conditions for its immediate reversal. A complete analysis of this complex system is something that we cannot offer. But the foregoing does offer some analytic insight into the contagious dynamics of equal division in "bargaining with neighbors."

---

[8] With some minor complications involving ties.

6. Conclusion

Sometimes we bargain with neighbors, sometimes with strangers. The dynamics of the two sorts of interaction are quite different. In the bargaining game considered here, bargaining with strangers—modeled by the replicator dynamics—leads to fair division from a randomly chosen starting point about 60 percent of the time. Fair division becomes the unique answer in bargaining with strangers if we change the question to that of stochastic stability in the ultra-long-run. But long expected waiting times call the explanatory significance of the stochastic stability result into question.

Bargaining with neighbors almost always converges to fair division and convergence is remarkably rapid. In bargaining with neighbors, the local interaction generates clusters of those strategies which are locally successful. Clustering and local interaction together produce positive correlation between like strategies. As noted elsewhere (Skyrms 1994, 1996), positive correlation favors fair division over the polymorphisms. In bargaining with neighbors, this positive correlation is not something externally imposed but rather an unavoidable consequence of the dynamics of local interaction. As a consequence, once a small group of demand-half players is formed, justice becomes contagious and rapidly takes over the entire population.

Both bargaining with strangers and bargaining with neighbors are artificial abstractions. In initial phases of human cultural evolution, bargaining with neighbors may be a closer approximation to the actual situation than bargaining with strangers. The dynamics of bargaining with neighbors strengthens the evolutionary explanation of the norm of fair division.

References

Anderlini, L. and A. Ianni (1997) "Learning on a Torus." In *The Dynamics of Norms*, ed. C. Bicchieri, R. Jeffrey, and B. Skyrms, 87–107. New York: Cambridge.

Axtell, R., J. M. Epstein, and H. P. Young (1999) "The Emergence of Economic Classes in an Agent-Based Bargaining Model." Preprint, Brookings Institution.

Björnerstedt, J. and J. Weibull (1996) "Nash Equilibrium and Evolution by Imitation." In *The Rational Foundations of Economic Behavior*, ed. Kenneth J. Arrow et al., pp. 155–71. New York: Macmillan.

Ellison, G. (1993) "Learning, Local Interaction and Coordination." *Econometrica* 61: 1047–71.

Eshel, I. E. Sansone, and A. Shaked (1996) "Evolutionary Dynamics of Populations with a Local Interaction Structure." Working paper, University of Bonn.

Foster, D. and H. P. Young (1990) "Stochastic Evolutionary Game Dynamics," *Theoretical Population Biology* 38: 219–32.

Hofbauer, J. and K. Sigmund (1988) *The Theory of Evolution and Dynamical Systems*. New York: Cambridge.

Lindgren, K. and M. Nordahl (1994) "Evolutionary Dynamics in Spatial Games." *Physica D* 75: 292–309.

Nash, J. (1950) "The Bargaining Problem." *Econometrica* 18: 155–62.

Nowak, M. A. and R. M. May (1992) "Evolutionary Games and Spatial Chaos." *Nature* 359: 826–29.

Nydegger, R. V. and G. Owen (1974) "Two-Person Bargaining: An Experimental Test of the Nash Axioms." *International Journal of Game Theory* 3: 239–50.

Pollack, G. B. (1989) "Evolutionary Stability on a Viscous Lattice." *Social Networks* 11: 175–212.

Roth, A. and M. Malouf (1979) "Game Theoretic Models and the Role of Information in Bargaining." *Psychological Review* 86: 574–94.

Samuelson, L. (1997) *Evolutionary Games and Equilibrium Selection*. Cambridge: MIT.

Schlag, K. (1996) "Why Imitate, and If So How?" Discussion Paper B-361, University of Bonn, Germany.

Skyrms, B. (1994) "Sex and Justice." *Journal of Philosophy* 91, 6: 305–20.

Skyrms, B. (1996) *Evolution of the Social Contract*. New York: Cambridge University Press.

Sugden, R. (1986) *The Economics of Rights, Cooperation, and Welfare*. New York: Blackwell.

Van Huyck, J., R. Batallio, S. Mathur, P. Van Huyck, and A. Ortmann (1995) "On the Origin of Convention: Evidence From Symmetric Bargaining Games." *International Journal of Game Theory* 24: 187–212.

Weibull, J. W. (1995) *Evolutionary Game Theory*. Cambridge: MIT.

Yaari, M. and M. Bar-Hillel (1981) "On Dividing Justly." *Social Choice and Welfare* 1: 1–24.

Young, H. P. (1993a) "An Evolutionary Model of Bargaining." *Journal of Economic Theory* 59: 145–68.

Young, H. P. (1993b) "The Evolution of Conventions." *Econometrica* 61: 57–94.

4

Stability and Explanatory Significance of Some Simple Evolutionary Models

1. Introduction

The explanatory value of equilibrium depends on the underlying dynamics. First there are questions of the dynamical stability of the equilibrium that are internal to the dynamical system in question. Is the equilibrium locally *stable*, so that states near to it stay near to it, or better, *asymptotically stable*, so that states near to it are carried to it by the dynamics? If not, we should not expect to see this equilibrium. But even if an equilibrium is asymptotically stable, that is no guarantee that the system will reach that equilibrium unless we know that the system's initial state is sufficiently close to the equilibrium. Global stability of an equilibrium, when we have it, gives the equilibrium a much more powerful explanatory role. An equilibrium is *globally asymptotically stable* if the dynamics carries every possible initial state in the interior of the state space to that equilibrium. If an equilibrium is globally stable, it can have explanatory value even when we are completely uncertain about the initial state of the system.

Once questions of dynamical stability are answered with respect to the dynamical system in question, there is the further question of the structural stability of that system itself. That is to say, are dynamical systems close to the one in question (in a sense to be made precise) topologically equivalent to that system? If not, a slight misspecification of the model may make predictions that are drastically wrong.

Structural stability is defined in terms of small changes in the model. But we may also be interested in what happens with some rather large changes in the model. A structurally stable model might, after all, be badly misspecified. Interest in such questions depends on the plausibility of the large changes being contemplated.

Here I would like to discuss these stability questions with respect to three simple dynamical evolutionary models from my book, *Evolution of the Social Contract* (1996). Two are models of simplified bargaining games, one with random encounters and one with correlated encounters. The third is a model of a simplified signaling game. These models all use replicator dynamics.

In modeling evolution—even cultural evolution—we face considerable uncertainty concerning early states of the system. Considerations of dynamical stability are therefore crucial to the evaluation of the explanatory significance of the model. If we can show only that an equilibrium is stable (that is to say, *locally* stable), we have only a "how possible" explanation. If we can show global stability, then the early state of the system is immaterial and, providing the dynamical system is the correct model, we have approached a "why necessarily" explanation.

Good reasons can be given why the replicator dynamics is a plausible candidate for a dynamics of cultural evolution. A number of different models of social learning by imitation have been shown to yield the replicator dynamics (Binmore, Gale, and Samuelson 1995; Sacco 1995; Björnerstedt and Weibull 1996; Schlag 1998). As I have said elsewhere, I believe that the replicator dynamics is a natural place to begin investigations of dynamical models of cultural evolution, but I do not believe that it is the whole story (Skyrms 1999). That means that structural stability, and more generally, stability under perturbations of the dynamics itself, are important. The more robust the result is to perturbations of the dynamics, the more likely it is to be of real significance in cultural evolution.

In *Evolution of the Social Contract* I make, without presenting proof, some claims about stability. I will substantiate those claims here. I will prove local asymptotic dynamic stability in the bargaining game with random encounters and global asymptotic dynamical stability in the other two models.

The dynamical stability results are of a rather different character for the bargaining game of chapter 1 (Skyrms 1996) than for the signaling game of chapter 5 (Skyrms 1996). Accordingly I make explanatory claims of different strengths for these two cases. In the bargaining game with random encounters, there are two attracting equilibria: one where everyone settles for equal shares and one where there is a population in which some demand a lot and some demand a little. I argue that, in the game in question, the equal division equilibrium is the one selected by commonly held norms of justice. The equal division equilibrium has the largest basin of attraction, but the basin of attraction of the inegalitarian equilibrium is not so small as to be negligible. At this point the power of the dynamics to explain cultural evolution of egalitarian norms is not impressive, and I say so. However, it may be that such norms evolved in an environment where encounters are not random, but there is positive correlation between types—for which various reasons might be given. A small amount of positive correlation incorporated in the second bargaining model eliminates the inegalitarian basin of attraction, and makes the equal division equilibrium a global attractor. (Larger amounts of positive correlation have the same qualitative result.) I conclude: "This is, perhaps, the beginning of an explanation of the origin of our concept of justice" (Skyrms 1996: 21). The words were chosen carefully. The claim of explanatory significance is meant to be modest.

In the case of the signaling game, the dynamical stability properties are quite different. In the model with random encounters and with one signaling system equilibrium and two anti-signaling system equilibria (my third model), the signaling system equilibrium is a global attractor. This remains true if positive correlation is added. In a larger model with two signaling system equilibria, almost all possible initial populations are carried to one signaling system equilibrium or another. In this case the dynamical stability results are much stronger than in the bargaining game, and I made a much stronger explanatory claim: "The emergence of meaning is a moral certainty" (Skyrms 1996: 93). (For discussion and qualification of this claim see Section 3.)

In a recent article, D'Arms, Batterman, and Górny, with whom I agree on the importance of stability, raise questions about the structural stability of my second model (D'Arms, Batterman, and Górny 1998: 91). In fact, I will show that this is the *only* one of the three models that *is* structurally stable. Modification of model 3 to allow correlated encounters, however, results in a structurally stable system. I will then show that a number of the foregoing results remain true if we substitute any dynamics in a large class of *qualitatively adaptive* dynamics. This strengthens the explanatory force of my models, because it identifies behavior that does not depend on the replicator dynamics or some dynamics very close to it being the right dynamics to use in a theory of cultural evolution. Of course, it is possible that dynamics that are even further afield may be of interest and importance. But I think that it is useful to have the questions addressed here settled. The techniques used to do so may also be of interest in the analysis of more complicated models.

2. The Three Dynamical Models

The dynamics underlying each of these models is the *replicator dynamics*. The proportion of the population playing strategy i is x_i. In each of the models considered here there are three strategies, so the state of the population is a vector, $\mathbf{x} = <x_1, x_2, x_3>$. The state space is the 3-simplex where $x_1 + x_2 + x_3 = 1$, with x_1, x_2, x_3, non-negative, as shown in Figure 4.1. The average fitness of strategy i in population state $\mathbf{x}$ is

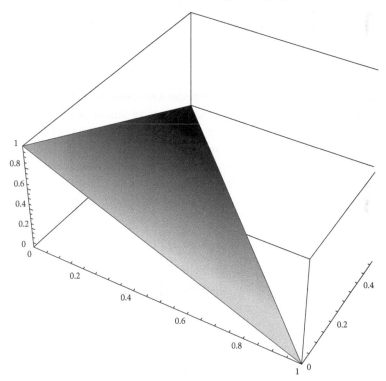

Figure 4.1 A three-simplex

denoted by "$U(x_i \mid \mathbf{x})$". [For future reference we also introduce $U(\mathbf{y} \mid \mathbf{x})$, the average fitness of an infinitesimal subpopulation playing strategies in the proportion specified by vector $\mathbf{y}$ when random paired with members of a population in states $\mathbf{x}$, as $\Sigma_i \, y_i \cdot U(x_i \mid \mathbf{x})$.] The average fitness of the whole population in state $\mathbf{x}$ is denoted by "$U(\mathbf{x} \mid \mathbf{x})$". The *replicator dynamics* is then given by the system of differential equations:

$$dx_i/dt = x_i \cdot [U(x_i \mid \mathbf{x}) - U(\mathbf{x} \mid \mathbf{x})]$$

The 3-simplex is invariant under this dynamics. If $\mathbf{x}$ is in the 3-simplex at a time, it remains within the 3-simplex for all time. The dynamics for three strategies thus lives on a plane.

The replicator dynamics was introduced by Taylor and Jonker (1978) as a simplified model of differential reproduction underlying the notion of evolutionarily stable strategy introduced by Maynard Smith and Price (1973). It has also been derived as the dynamics of various models of cultural evolution (Binmore, Gale, and Samuelson 1995; Björnerstedt and Weibull 1996; Schlag 1998) and as a limiting case of reinforcement learning (Borgers and Sarin 1997).

Model 1

The three models differ in how fitnesses are determined. In the first model individuals are paired at random from an infinite population to play a bargaining game. The three strategies are S_1: Demand 1/3 of the cake, S_2: Demand 2/3 of the cake, and S_3: Demand 1/2 of the cake. If demands total more that 1, no one gets anything; otherwise, players get what they demand. Fitness here equals amount of cake. Those who demand 1/3 have demands that are compatible with all players and always get 1/3:

$$U(x_1 \mid \mathbf{x}) = 1/3$$

Those who demand 2/3 only get their demand when they are paired with those who demand 1/3; otherwise they get nothing. Their average fitness is:

$$U(x_2 \mid \mathbf{x}) = 2/3 \cdot x_1$$

Those who demand 1/2 get their demand except when matched against those who demand 2/3. Their average fitness is:

$$U(x_3 \mid \mathbf{x}) = 1/2 \cdot (1 - x_2)$$

The average fitness of the population is gotten by averaging the average fitnesses of the various strategies:

$$U(\mathbf{x} \mid \mathbf{x}) = x_1 \cdot U(x_1 \mid \mathbf{x}) + x_2 \cdot U(x_2 \mid \mathbf{x}) + x_3 \cdot U(x_3 \mid \mathbf{x})$$

Model 2

In the second model the strategies and basic game are the same as in model 1, but the individuals are not paired at random. Rather, there is some positive correlation in the encounters determined by a parameter, e. The probability that strategy i meets itself, $p(S_i|S_i)$ is not simply the proportion of the population playing that strategy, x_i, as in the random pairing model, but rather it is inflated thus:

$$p(S_i|S_i) = x_i + e \cdot (1 - x_i)$$

The probability of strategy S_i meeting a different strategy, S_j is correspondingly deflated:

$$p(S_j|S_i) = x_j - e \cdot x_j$$

We will take $e = 1/5$, which is a case discussed both in Chapter 1 of my book (1996) and in D'Arms, Batterman, and Górny (1998). Then, as before:

$$U(x_1 \mid \mathbf{x}) = 1/3$$

but:

$$U(x_2 \mid \mathbf{x}) = 2/3 \cdot 4/5 \cdot x_1$$

and:

$$U(x_3 \mid \mathbf{x}) = 1/2 \cdot (x_3 + 1/5 \, (1 - x_3)) + 1/2 \cdot 4/5 \cdot x_1$$

The average fitness of the population is calculated as before.

Model 3

This is a random pairing model like model 1, but the underlying game is a signaling game from Chapter 5, pp. 91–3, of my book (1996). There are two antisignaling system strategies, x_1, x_2, and one signaling system strategy, x_3. Each antisignaling system strategy does badly against itself for a fitness of zero, well against the other for a fitness of one and middling against the signaling system for fitness of one-half. The signaling system has fitness one-half against either of the antisignaling systems and fitness of one against itself. This gives the following average fitnesses for the three strategies:

$$U(x_1 \mid \mathbf{x}) = x_2 + (1/2) \cdot x_3$$
$$U(x_2 \mid \mathbf{x}) = x_1 + (1/2) \cdot x_3$$
$$U(x_3 \mid \mathbf{x}) = x_3 + 1/2 \cdot (x_1 + x_2)$$

3. Local Dynamical Stability of Equilibria

First we will identify the equilibria and their local stability characteristics in each of our three models. The equilibria can be identified by solving equations. Their dynamical stability characteristics can be investigated by evaluating the eigenvalues of the Jacobian matrix of partial derivatives at the equilibrium point (see, e.g., Hirsch and Smale 1974: ch. 9). If the eigenvalues all have non-zero real part, the equilibrium is said to be *hyperbolic* and the eigenvalues determine the local dynamical stability properties of the equilibrium. If the real parts of the eigenvalues are all negative, then the equilibrium is called a *sink*, and it is asymptotically stable. If the real parts of the eigenvalues are a positive, the equilibrium is called a *source*, and if both positive and negative real parts occur it is called a *saddle*. Sources and saddles are unstable. If the point is non-hyperbolic, local stability must be investigated by different means.

Model 1

Recall that this is the bargaining game with just three strategies, S_1 (Modest) = Demand 1/3; S_2 (Greedy) = Demand 2/3; S_3(Fair) = Demand 1/2. Each vertex of the simplex represents a state in which the population is totally composed of one of the types. Here, as always in the replicator dynamics, each vertex is a dynamic equilibrium. (If the other guys are extinct, they cannot reproduce.) At an equilibrium in the interior of the S_1–S_2 edge, both these strategies must have the same fitness. Solving the equations: $x_1 + x_2 = 1$ and $(1/3) = (2/3) x_1$, we find a Modest–Greedy equilibrium at $x_1 = 1/2$, $x_2 = 1/2$. There are no equilibria interior to the other edges. An equilibrium interior to the simplex must have all three strategies having equal fitnesses. Solving these equations we find a Modest–Greedy–Fair equilibrium at $x_1 = 1/2$, $x_2 = 1/3$, $x_3 = 1/6$. These five states are the only equilibrium states in this model.

We proceed to examine the eigenvalues of the Jacobian at these points. Since the state space—the three-simplex—is a two-dimensional object, we need to consider the dynamics in terms of only two of the three variables; any two will do. For each point we will get two eigenvalues. (We could evaluate eigenvalues for the Jacobian in the three variable system, but this would generate one spurious zero eigenvalue, associated with the eigenvector pointing out of the simplex (see Bomze 1986, 48; or van Damme 1987, 222).) All the calculations of eigenvalues reported in this paper were performed using *Mathematica*. The calculations for Model 1 are given in the appendix.

At the All Fair equilibrium, $x_3 = 1$ the eigenvalues are $\{-1/2, -1/6\}$. The two negative eigenvalues identify this as a *sink*—an *asymptotically stable* equilibrium. The Modest–Greedy equilibrium at $x_1 = x_2 = 1/2$ is likewise *asymptotically stable* with a pair of negative eigenvalues $\{-1/6, -1/12\}$. The Modest–Greedy–Fair equilibrium in the interior of the simplex at $x_1 = 1/2$, $x_2 = 1/3$, $x_3 = 1/6$, has one negative and one positive eigenvalue. The values are approximately $\{-0.146525, 0.061921\}$. This indicates that this equilibrium is a *saddle*, which is attracting in one direction and repelling in another. This equilibrium is dynamically unstable. The All Modest equilibrium at $x_1 = 1$ has two positive eigenvalues $\{1/6, 1/3\}$, which identifies it as a *source*, an unstable repelling equilibrium. This leaves All Greedy equilibrium at $x_2 = 1$. This has eigenvalues $\{0, 1/3\}$. Because of the zero eigenvalue, this is not a hyperbolic equilibrium, and its local dynamical stability properties cannot be completely inferred from the eigenvalues of the Jacobian. The one positive eigenvalue, however, shows that it is not stable. We will be able to say more about it, using different techniques, in the next section. The information that we have about model 1 so far is summarized in Table 4.1.

Model 2

Model 2 is the bargaining game with correlation of encounters at the .2 level. There are four equilibria. Each of the vertices is an equilibrium. There is an equilibrium on the Greedy–Modest edge, at $x_1 = 5/8$, $x_2 = 3/8$. There is no equilibrium in the interior of the 3-simplex. When we

Table 4.1 Stability of equilibria for Model 1

Equilibrium	Eigenvalues	Stability
$x_1 = 1, x_2 = 0, x_3 = 0$	1/6, 1/3	Unstable (source)
$x_1 = 0, x_2 = 1, x_3 = 0$	0, 1/3	(non-hyperbolic)
$x_1 = 0, x_2 = 0, x_3 = 1$ (All Fair)	$-1/2, -1/6$	Stable (sink)
$x_1 = 1/2, x_2 = 1/2, x_3 = 0$	$-1/6, -1/12$	Stable (sink)
$x_1 = 1/2, x_2 = 1/3, x_3 = 1/6$	$-0.146525, 0.0631921$	Unstable (saddle)

Table 4.2 Stability of equilibria for Model 2

Equilibrium	Eigenvalues	Stability
$x_1 = 1, x_2 = 0, x_3 = 0$	1/6, 1/5	Unstable(source)
$x_1 = 0, x_2 = 1, x_3 = 0$	1/10, 1/3	Unstable (source)
$x_1 = 0, x_2 = 0, x_3 = 1$ (All Fair)	$-1/2, -1/6$	Stable (sink)
$x_1 = 5/8, x_2 = 3/8, x_3 = 0$	$-1/8, 1/60$	Unstable (saddle)

evaluate the eigenvalues of the Jacobian at these equilibria we find that they are all hyperbolic, so we have a complete characterization of the local stability characteristics of these equilibria. These are collected in Table 4.2.

Model 3

In model 3, x_1 and x_2 are two "antisignaling system strategies" and x_3 is a signaling system strategy. There are four equilibria—the three vertices and an antisignaling polymorphism at $x_1 = x_2 = 1/2$. We have three hyperbolic equilibria and one non-hyperbolic equilibrium. The analysis is summarized in Table 4.3.

Table 4.3 Stability of equilibria for Model 3

Equilibrium	Eigenvalues	Stability
$x_1 = 1, x_2 = 0, x_3 = 0$	1/2, 1	Unstable (source)
$x_1 = 0, x_2 = 1, x_3 = 0$	1/2, 1	Unstable (source)
$x_1 = 0, x_2 = 0, x_3 = 1$ (Signaling)	$-1/2, -1/2$	Stable (sink)
$x_1 = 1/2, x_2 = 1/2, x_3 = 0$	$-1/2, 0$	(non-hyperbolic)

4. Global Dynamical Stability of Equilibria

The sinks identified in Section 3 are dynamically asymptotically stable. This is a local property that is established by dynamical behavior in some neighborhood of the equilibrium. But the neighborhood might be very small. It would be more powerful if we could demonstrate that an equilibrium has a significant basin of attraction, or even that it is globally asymptotically stable—that the dynamics carries every point in the interior of the state space to it, in the limit.

The main technique employed in this section is the use of the Kullback–Leibler relative entropy as a Liapunov Function for replicator dynamics. A Liapunov function is a generalization of the notion of potential. Liapunov showed that if x is an equilibrium and V is a continuous real-valued function defined on some neighborhood, W, of x, differentiable on W-x such that:

(i) $V(x) = 0$ and for $y \neq x$ in W, $V(y) > 0$.
(ii) The time derivative of V, $V' < 0$ in W-x,

then the orbit of any point in W approaches x as time goes to infinity. If these requirements can be shown to hold taking the neighborhood as the whole state space (or its interior), then x is *globally asymptotically stable* (see Hirsch and Smale 1974: ch. 9, § 3; and Guckenheimer and Holmes 1986: 5ff.).

The Kullback–Leibler relative entropy serves as an appropriate Liapunov function. The Kullback-Leibler relative entropy of state y with respect to state x is:

$$H_x(y) = \Sigma_i \, x_i \, \log(x_i/y_i)$$

with the sum being taken over the carrier of x, that is to say over the strategies that have a positive population proportion in state x. This function meets the requirements of continuity and differentiability for a Liapunov function and assumes its minimum at x.

Its time derivative of $H_x(y)$ is just $-[U(x \mid y) - U(y \mid y)]$, which we need to be negative to complete the requirements for a Liapunov function. So we only need to check that $[U(x \mid y) - U(y \mid y)]$ is positive throughout the neighborhood in question (see Bomze 1991; Weibull 1997: 3.5 and 6.5).

Model 3

We can now prove global convergence to the signaling system equilibrium in the signaling game of model 3. The state of fixation of the signaling system strategy is at $x_3 = 1$. We consider the neighborhood, $x_3 > 0$, where this strategy is not extinct. This includes the vertex, $x_3 = 1$, the $x_1 - x_3$ and $x_2 - x_3$ edges, and the interior of the 3-simplex. We use as a Liapunov function the entropy relative to the state where $x_3 = 1$. We now need to check that if $x_3 > 0$, $U(x_3 \mid \mathbf{x}) - U(\mathbf{x} \mid \mathbf{x})$ is non-negative everywhere and equal to zero only where $x_3 = 1$.

First, we want to show that for a fixed value of x_3, $U(x_3 \mid \mathbf{x}) - U(\mathbf{x} \mid \mathbf{x})$ assumes its minimum at the point where $x_1 = x_2$. Then we need only check that on the line, $x_1 = x_2$, to establish the desired result. For a fixed value of x_3, $U(x_3 \mid \mathbf{x}) = 1/2\, x_1 + 1/2\, x_2 + x_3$ is constant so $U(x_3 \mid \mathbf{x}) - U(\mathbf{x} \mid \mathbf{x})$ assumes its minimum when $U(\mathbf{x} \mid \mathbf{x})$ assumes its maximum. For fixed x_3, $(\mathbf{x} \mid \mathbf{x}) = x_1\, U(x_1) + x_2\, U(x_2) + x_3\, U(x_3)$ assumes its maximum when $x_1\, U(x_1) + x_2\, U(x_2)$ does. The quantity, $x_1\, U(x_1) + x_2\, U(x_2) = x_1(x_2 + x_3/2) + x_2(x_1 + x_3/2) = 2x_{12} + x_3/2(x_1 + x_2)$ now assumes its maximum where $2x_1x_2$ does, that is where $x_1 = x_2$.

Now consider a point, $\mathbf{x}$, on the line, $x_1 = x_2$. Here $U(x_3 \mid \mathbf{x}) = (1/2)(2x_1) + (1 - 2x_1) = 1 - x_1$. And $U(x_2 \mid \mathbf{x}) = U(x_1 \mid \mathbf{x}) = x_2 + (1/2)\, x_3 = x_1 + (1/2)(1 - 2\, x_1) = 1/2$. So when x_1, x_2, x_3, are all positive, $U(x_3 \mid \mathbf{x}) > U(\mathbf{x} \mid \mathbf{x})$ when $(1 - x_1) > 1/2$. On the line $x_1 = x_2$ in the interior of the simplex, $1/2 > x_1 > 0$ so, $U(x_3 \mid \mathbf{x}) > U(\mathbf{x} \mid \mathbf{x})$. By definition at $x_3 = 1$, $U(x_3 \mid \mathbf{x}) = U(\mathbf{x} \mid \mathbf{x})$. Thus, in the global neighborhood, $x_3 > 0$, where the signaling strategy is not extinct, our Liapunov function is non-negative and equal to zero only at the point of fixation of that strategy. In Model 3 we have global convergence to the signaling system equilibrium.

This result answers the lingering question about the local stability properties of the equilibrium $x_1 = x_2 = 1/2$, that was left by the zero eigenvalue that showed up in the local stability analysis of the previous section. This equilibrium is locally dynamically unstable within the 3-simplex. However, within the subsimplex consisting of the $x_1 - x_2$ line (where x_3 is extinct) this equilibrium is globally stable. This can be shown by applying the same techniques within the subsimplex.

Model 2

The qualitative analysis of Model 2 is much like that of model 3. We use the same relative entropy Liapunov function. The quantity

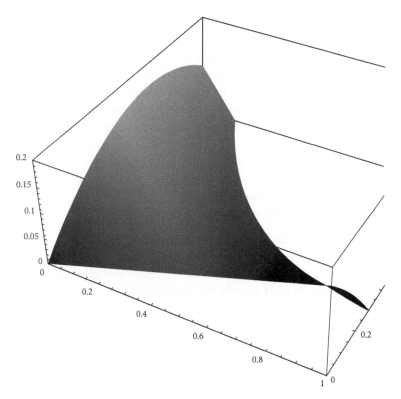

Figure 4.2 Liapunov Function for Model 2

$U(x_3 \mid \mathbf{x}) - U(\mathbf{x} \mid \mathbf{x})$, which simplifies to $1/30\ (x_1\ (5 - 28\ x_2) + 3(5 - 4\ x_2)\ x_2)$, is positive throughout the entire region where $x_3 > 0$. Here I will let a picture stand in for the algebraic proof. Figure 4.2 is a plot of this function above the $x_1 - x_2$ plane. It is positive on the relevant region $[x_1 > 0, x_2 > 0, (x_1 + x_2) < 1)]$. The line $x_3 = 0$, is a subsimplex. On this subsimplex the Greedy–Modest mixed equilibrium is global asymptotically stable (even though it is not even locally stable on the whole 3-simplex).

Model 1

In model 1, there is no globally asymptotically stable equilibrium. Both the All Fair equilibrium, $x_3 = 1$, and the Greedy–Modest equilibrium, $x_2 = x_3 = 1/2$, are asymptotically stable sinks. They both have basins of attraction that include points in the interior of the 3-simplex. Liapunov functions can be used as before, however, to prove that an equilibrium attracts all points in some significantly extended neighborhood. As an illustration, consider the neighborhood of the All Fair equilibrium defined by $x_3 > .7$. (This is by no means the whole basin of attraction for this equilibrium, but it makes for a quick example.) It is easy to see that $U(x_3 \mid x) - U(x \mid x)$ must be positive throughout this neighborhood. The minimum value that $U(x_3 \mid x)$ can have is $(.7)(.5) = .35$. $U(x_1 \mid x)$ is constant at $1/3$. $U(x_2 \mid x)$ has its maximum at $(2/3)(.3) = .2$. So $U(x_3 \mid x)$ must always be greater than the average, $U(x \mid x)$. We will have reason refer back to this illustration in Section 6.

To summarize, we have established:

4.1 The All Fair and Signaling System equilibria, of models 2 and 3 respectively, are *globally asymptotically stable*.

4.2 The All Fair equilibrium in model 1 is *locally asymptotically stable* and attracts every point in the neighborhood $x_3 > .7$.

5. Structural Stability of the Dynamical System

A dynamical system is *structurally stable* if small enough, but otherwise arbitrary, perturbations result in a topologically equivalent system. That is to say that all dynamical systems sufficiently close to the one under consideration, are topologically equivalent. This will be made precise shortly.

It is sometimes held that only structurally stable dynamical systems have explanatory value, but this *stability dogma*, has also been questioned (as in Guckenheimer and Holmes 1986: ch. 5). The point is that the only perturbations that are crucial are those that are physically possibly for the phenomena under consideration. To demand structural stability may be to demand too much. It might also be to demand too little, as plausible perturbations of the dynamical system might be large.

Nevertheless, is may be useful to see whether a dynamical system is structurally stable, and if it fails to be so to see how it fails. With regard to the models under consideration here, the question of structural stability of model 2 is raised by D'Arms, Batterman, and Górny (1998: 91). (They also have concerns about larger perturbations of the model.) It may be somewhat surprising, then, that we will be able to prove that of our three models only Model 2 is structurally stable.

First, we need a precise definition of structural stability (Piexoto 1962; Smale 1980; Guckenheimer and Holmes 1986). In each of our models, the differential equations of the replicator dynamics generate a dynamical system on the compact subset of the $x_1 - x_2$ plane, where x_1 and x_2 are non-negative and their sum does not exceed one. Call this region, M. (As noted earlier, we need only consider this system, since the values of x_1 and x_2 determine the value of x_3.) The planar nature of M simplifies the following discussion in a number of ways.

The differential equations define a continuously differentiable *vector field* on this region, M. The vector field is a function that associates with every point the vector $\langle dx_1/dt, dx_2/dt \rangle$ with the derivatives being evaluated at that point. This vector field can be thought of as being the dynamical system. It determines a family of *solution curves*, or *orbits*, on M. Two dynamical systems are *topologically equivalent* if there is a homeomorphism taking from M to M, which preserves orbits and their temporal sense.

To consider dynamical systems "close" to one of our models, we need a space of dynamical systems and a sense of closeness. For the space of dynamical systems, we take the continuously differentiable vector fields on M, $\chi(M)$. Closeness can be defined relatively simply because of the nice nature of our underlying state space, M. For each dynamical system, X in $\chi(M)$, let its norm, $\|X\|$, be that maximum of the following numbers:

least upper bound $|\,dx_1/dt\,|$
least upper bound $|\,dx_2/dt\,|$
least upper bound $|\,\partial[dx_1/dt]/\partial x_1\,|$
least upper bound $|\,\partial[dx_1/dt]/\partial x_2\,|$
least upper bound $|\,\partial[dx_2/dt]/\partial x_1\,|$
least upper bound $|\,\partial[dx_2/dt]/\partial x_2\,|$

with the least upper bounds being taken over all points in M. Now we can define a *metric* on the space of dynamical systems, $\chi(M)$. For two such dynamical systems, X, Y, we take $d(x,y) = \|X - Y\|$. Now we can give a precise definition of structural stability:

A vector field, X, is *structurally stable*, if there is a neighborhood of X such that every vector field, Y, in that neighborhood is *topologically equivalent* to X.

The tool to be used in this section to establish structural stability is Peixoto's Theorem (Peixoto 1962; Guckenheimer and Holmes 1986: 60): *A vector field defined on a compact region of the plane is structurally stable **if and only if***

(1) The number of equilibrium points and closed orbits is finite and each is hyperbolic.
(2) There are no orbits connecting saddle points.

(We note again that the planar nature of M simplifies the form of the theorem used here.)

Since the conditions are necessary for structural stability, and since models 1 and 3 have non-hyperbolic equilibria, we can conclude that these dynamical systems are not structurally stable. What about model 2? In Section 3, we saw that it has a finite number of equilibria, all of which are hyperbolic. In Section 4 we saw that the "All Fair" equilibrium at $x_3 = 1$ is globally asymptotically stable for the sub-region where $x_3 > 0$. The leaves only the line $x_3 = 0$. We saw that the Greedy–Modest mixed equilibrium is globally asymptotically stable on the interior of that line. Therefore there are no closed orbits and there are no orbits connecting saddles. Since the conditions are sufficient for structural stability, model 2 is structurally stable.

5.1 Models 1 and 3 are not structurally stable
5.2 Model 2 is structurally stable.

(Analysis of structural stability is more complicated in higher dimensional systems. Model 2 is a Morse–Smale system (Guckenheimer and Holmes 1986: 64). In higher dimensions being Morse–Smale is a sufficient condition for structural stability, but is no longer a necessary condition.)

6. Qualitatively Adaptive Dynamics

Let us say that a dynamics is *qualitatively adaptive* if, according to that dynamics, a strategy that is not extinct increases its proportion of the population if its fitness is higher that the population average, decreases its population proportion if its fitness is less that the population average, and keeps the same population proportion if its fitness is equal to the population average. That is to say:

If $x_i > 0$, then dx_i/dt agrees in sign with $U(x_i \mid x) - U(x \mid x)$

The replicator dynamics is a member of this class, but there are many other qualitatively adaptive dynamics. Substituting another qualitatively adaptive dynamics for the replicator dynamics in our three models can significantly perturb the vector field. Structural stability does not guarantee anything about behavior of the dynamical systems under this class of perturbations. Nevertheless we can show that the global stability results of Section 4 generalize to all these systems.

We use the same relative entropy as a Liapunov function. To prove convergence to the equilibrium as $x_3 = 1$, the entropy relative to the equilibrium at state x is just:

$$-\log (x_3)$$

This function is continuous and non-negative on the simplex and equals zero only at the equilibrium. Its time derivative is:

$$-(1/x_3) \ dx_3/dt$$

To prove convergence, we now need only to show that this is negative on the region consisting of the neighborhood $x_3 > 0$ with the equilibrium removed—that its to say when $0 < x_3 < 1$. Here the time derivative is negative just where dx_3/dt is positive, which is where $U(x_3 \mid x) - U(x \mid x)$ is positive since the dynamics is adaptive.

Thus the results of Section 4 carry over:

6.1 With any *qualitatively adaptive dynamics* substituted for the replicator dynamics, the All Fair and Signaling System equilibria, of models 2 and 3 respectively, are *globally asymptotically stable*.

6.2 For any *qualitatively adaptive dynamics* substituted for the replicator dynamics, the All Fair equilibrium in model 1 is *asymptotically stable* and attracts every point in the neighborhood $x_3 > .7$.

We can even strengthen (6.2) to include a broader class of dynamics. Say that a dynamics is *weakly qualitatively adaptive* if a strategy, s_i, has positive growth in population proportion, $dx_i/dt > 0$, in any state in which it is not extinct and its fitness is the highest of any strategy represented in the population. (A dynamics may be weakly qualitatively adaptive, without being qualitatively adaptive if, for instance, being second best and better than the average counts for nothing.) Now we say that in model 1, if $x_3 > .7$, then strategy 3, Demand 1/2, is the fittest strategy. Then by the same reasoning as before we have:

> **6.3** For any *weakly qualitatively adaptive dynamics* substituted for the replicator dynamics, the All Fair equilibrium in model 1 is *asymptotically stable* and attracts every point in the neighborhood $x_3 > .7$.

In Chapter 1 of *Evolution of the Social Contract*, in a passage quoted by D'Arms, Batterman, and Górny, I claimed that the fact that the All Fair equilibrium is asymptotically stable was robust for a wide class of adaptive dynamics:

[Demand 1/2's] strong stability properties guarantee that it is an attracting equilibrium in the replicator dynamics, but also make the details of that dynamics unimportant. Fair division will be stable in any dynamics with a tendency to increase the proportion (or probability) of strategies with greater payoffs, because any unilateral deviation from fair division results in a strictly worse payoff. For this reason, the Darwinian story can be transposed into the context of *cultural evolution*, in which imitation and learning play an important role in the dynamics. (Skyrms 1996: 11)

Proposition **6.2** establishes the technical claim that I make in this passage, and proposition **6.3** strengthens it in a way that supports my general conclusion (Skyrms 1996, 11).

7. Correlation Structure

Is model 2 robust against arbitrary changes in the correlation structure? Model 1 shows that it is not. Model 2 is not topologically equivalent to model 1. If you take model 2 and gradually reduce the correlation you come to a point where there is a bifurcation and the four equilibrium points of model 2 become the five equilibrium points of model 1. The Modest–Greedy equilibrium changes from a saddle to a sink. Arbitrary changes to the correlation structure can make a qualitative difference here, as they can in almost any game. (D'Arms, Batterman, and Górny make the same point by constructing an alternative model with anti-correlation.) But we know that model 2 is robust against local changes in correlation structure, those within some small neighborhood, as a result of our structural stability result.

Model 3 is not structurally stable. What happens to it if we add a small amount of positive correlation? We get model 4.

Model 4

$$U(x_1 \mid \mathbf{x}) = (1 - e) \cdot x_2 + (1/2) \cdot (1 - e) \cdot x_3$$
$$U(x_2 \mid \mathbf{x}) = (1 - e) \cdot x_1 + (1/2) \cdot (1 - e) \cdot x_3$$
$$U(x_3 \mid \mathbf{x}) = x_3 + e \cdot (x_1 + x_2) + 1/2 \cdot (1 - e) \cdot (x_1 + x_2)$$

With e = 0, model 4 reduces to model 3 and the equilibrium at $x_1 = x_2 = 1/2$ is non-hyperbolic with eigenvalues $\{-1/2, 0\}$. In model 4, the eigenvalues of the Jacobian evaluated at this equilibrium are:

$$\{1/4(-1 + 3e - SQR(1 + 2e + e^2), 1/4(-1 + 3e + SQR(1 + 2e + e^2)\}$$

For any small positive e, the equilibrium becomes hyperbolic. For example, for e = .001, the eigenvalues are $\{-0.4995, 0.001\}$. The equilibrium is a saddle, and thus unstable. The local stability properties of the other equilibria remain unchanged. We now have a *structurally* stable system that bears a qualitative resemblance to model 2. (That is to say that there are two sources at $x_1 = 1$ and at $x_2 = 1$, a mixed equilibrium saddle on the $x_1 - x_2$ line, and a sink at $x_3 = 1$.)

Clearly, correlation structure can be very important in determining the stability properties of an evolutionary dynamical system. It is of particular interest to consider alternative models where correlation is endogenously generated. D'Arms, Batterman, and Górny (1998) have ideas in this direction that might be profitably pursued in more detail. If I am not mistaken, with easier (and therefore higher levels of) correlation and anti-correlation of the kind that they consider in their paper, the Greedy–Modest equilibrium again becomes unstable. It would be interesting to have the whole story.

An alternative route to endogenous correlation structure uses interactions with neighbors in a spatially explicit structure. In Alexander (1999) and Alexander and Skyrms (1999) it is shown that the correlation that emerges in this sort of model for bargaining games (like that of model one) has dramatic effects on both the proportion of initial conditions that evolve to the All Fair equilibrium, and the speed with which that process takes place.

8. Stability and Explanatory Significance

Stability considerations are only one ingredient to be taken into account in assessing the explanatory significance of dynamical models. There are also larger considerations regarding the aptness of the model (or class of models) to the process being modeled—in this case the process of cultural evolution. I will not address these larger considerations here. They have been discussed elsewhere (D'Arms 2000; Barrett et al. 1999; Bicchieri 1999; Bolton 1997, 2000; Carpenter 2000; Gintis 2000; Güth and Güth 2000; Harms 2000; Kitcher 1999; Krebs 2000; Mar 2000; Nesse 2000; Proulx 2000; Skyrms 1999, 2000). But, bracketing these concerns, we can ask what bearing the stability results of this paper have on the explanatory significance of the models discussed here.

The fact that the All Fair equilibrium is an attractor in the bargaining game of models 1 and 2 is extremely robust. It holds not only in both these models, but also in models in which the replicator dynamics is replaced by any other dynamics in the class of weakly qualitatively adaptive dynamics. But the All Fair equilibrium is not a global attractor in model 1. Adding a modest amount of positive correlation (e = 1/5 or more) turns it into one, in a structurally stable dynamical system. Independent arguments for this degree of positive correlation during the evolution of the norm (or for some modified dynamical model) would be required to complete the explanation. My models do not give the whole story of the evolution of the equal split—they begin that story.

The fact that the signaling system equilibrium is a global attractor in model 3 and remains so if the replicator dynamics is replaced by any qualitatively adaptive dynamics, strikes me as explanatorily powerful. But the fact that model 3 is not structurally stable demands attention. Adding positive correlation takes us to model 4, which is structurally stable, and in which the signaling system equilibrium remains a global attractor. Here—in contrast to the case of the bargaining game—any amount of positive correlation, no matter how small, will do.

In general, these results should focus our attention on correlation. Correlation structure is of the utmost importance. Models of the co-evolution of strategic interaction with correlation structure are a promising direction for future research.

Acknowledgments

I would like to thank Bruce Bennett for a tutorial on structural stability and Morse–Smale systems, and two anonymous referees for helpful suggestions.

Appendix

Calculation of Eigenvalues of the Jacobian at Equilibria for Model 1 using Mathematica.

$In[1]:$ = Ux[x_, y_] : = (1/3);

Uy[x_, y_] : = (2/3) *x;

Uz[x_, y_] : = (1/2) * (1 − y);

Ubar[x_, y_] : = x^* Ux[x, y] + y^* Uy[x, y] + (1 − x − y) * Uz[x, y];

dxdt[x_, y_] : = x* (Ux[x, y] − Ubar[x, y]);

dydt[x_, y_] : = y* (Uy[x, y] − Ubar[x, y]);

$In[2]:$ = Outer[D, {dxdt[x, y], dydt[x, y]}, {x, y}]

$$Out[2]:= \left(\left(\left(\frac{1-y}{2}-\frac{1}{3}-\frac{2y}{3}\right)x - \frac{x}{3} - \frac{2xy}{3} - \frac{1}{2}(1-y)(-x-y+1)\right.\right.$$
$$\left. + \frac{1}{3}\left(\frac{1-y}{2} - \frac{2y}{3} + \frac{1}{3}\right)y\right)$$
$$x\left(-\frac{1}{3}(2x) + \frac{1-y}{2} + \frac{1}{2}(-x-y+1)\right)\frac{x}{3} - \frac{2xy}{3} - \frac{1}{2}(1-y)(-x-y+1)$$
$$\left.+ \left(-\frac{1}{3}(2x) + \frac{1-y}{2} + \frac{1}{2}(-x-y+1)y\right)\right)$$

$In[3]:$ = Eigenvalues [%]

$$Out[3] = \left\{ \frac{1}{12}(18y+7x-28xy-12y^2 - \right.$$
$$\sqrt{36y^4+168xy^3+196x^2y^2+60y^2+156xy+9x^2-12x-24y-324xy^2-72y^3+4}-4),$$
$$\frac{1}{12}(18y+7x-28xy-12y^2 -$$
$$\left.\sqrt{36y^4+168xy^3+196x^2y^2+60y^2+156xy+9x^2-12x-24y-324xy^2-72y^3+4}-4,)\right\}$$

$In[4]:$ = eig = %;

$In[5]:$ = eig / . {x → 1, y → 0}

$$Out[5]:=\left\{\frac{1}{6},\frac{1}{3}\right\}$$

In[6]: = eig / . {x → 0, y → 0}

$$Out[6]=\left\{-\frac{1}{2},-\frac{1}{6}\right\}$$

In[7]: = eig / . {x → (1/2), y → (1/2)}

$$Out[7]=\left\{-\frac{1}{6},-\frac{1}{12}\right\}$$

In[8]: = eig / . {x → (1/2), y → (1/3)}

$$Out[8]=\left\{\frac{1}{12}\left(-\frac{1}{2}-\frac{\sqrt{\frac{19}{3}}}{2}\right),\frac{1}{12}\left(-\frac{1}{2}+\frac{\sqrt{\frac{19}{3}}}{2}\right)\right\}$$

In[9]: = N[%]
Out[9] = {−0.146525, 0.0631921}
In[10]: = eig / . {x → 1, y → 0}

$$Out[10]=\left\{\frac{1}{6},\frac{1}{3}\right\}$$

In[11]: = eig / . {x → 0, y → 1}

$$Out[11]=\left\{0,\frac{1}{3}\right\}$$

References

Alexander, J. (1999) "The (Spatial) Evolution of the Equal Split", Institute for Mathematical Behavioral Science, University of California Irvine.

Alexander, J. and B. Skyrms (1999) "Bargaining with Neighbors: Is Justice Contagious?" *Journal of Philosophy* 96: 588–98.

Andronov, A. A. et al. (1971) *Theory of Bifurcations of Dynamical Systems on a Plane* (tr. from the Russian original of 1967). Jerusalem: Israel Program of Scientific Translations.

Barrett, M., E. Eells, B. Fitelson, and E. Sober (1999) "Models and Reality: A Review of Brian Skyrms's *Evolution of the Social Contract*." *Philosophy and Phenomenological Research* 59: 237–41.

Bicchieri, C. (1999) "Local Fairness." *Philosophy and Phenomenological Research* 59: 229–36.

Binmore, K., J. Gale, and L. Samuelson (1995) "Learning to be Imperfect: The Ultimatum Game." *Games and Economic Behavior* 8: 56–90.

Björnerstedt, J. and J. Weibull (1996) "Nash Equilibrium and Evolution by Imitation." In *The Rational Foundations of Economic Behavior*, ed. K. Arrow et al., 155–71. New York: Macmillan.

Bolton, G. (1997) "The Rationality of Splitting Equally." *Journal of Economic Behavior and Organization* 32: 365–81.

Bolton, G. (2000) "Motivation and the Games People Play." *Journal of Consciousness Studies* 7: 285–90.

Bomze, I. (1986) "Non-Cooperative Two Person Games in Biology: A Classification." *International Journal of Game Theory* 15: 31–59.

Bomze, I. (1991) "Cross-Entropy Minimization in Uninvadable States of Complex Populations." *Journal of Mathematical Biology* 30: 73–87.

Borgers, T. and R. Sarin (1997) "Learning Through Reinforcement and the Replicator Dynamics." *Journal of Economic Theory* 77: 1–14.

Carpenter, J. (2000) "Blurring the Line Between Rationality and Evolution." *Journal of Consciousness Studies* 7: 291–5.

D'Arms, J. (1996) "Sex, Fairness and the Theory of Games." *Journal of Philosophy* 96: 615–727.

D'Arms, J. (2000) "When Evolutionary Game Theory Explains Morality, What Does It Explain?", *Journal of Consciousness Studies* 7: 296–9.

D'Arms, J., R. Batterman, and K. Górny (1998) "Game Theoretic Explanations and the Evolution of Justice." *Philosophy of Science* 65: 76–102.

Gintis, H. (2000) "Classical vs. Evolutionary Game Theory." *Journal of Consciousness Studies* 7: 300–4.

Guckenheimer, J. and P. Holmes (1986) *Nonlinear Oscillations, Dynamical Systems, and Bifurcations of Vector Fields*. New York: Springer.

Güth, S. and W. Güth (2000) "Rational Deliberation versus Behavioral Adaptation: Theoretical Perspectives and Experimental Evidence." *Journal of Consciousness Studies* 7: 305–8.

Harms, W. (2000) "The Evolution of Cooperation in Hostile Environments." *Journal of Consciousness Studies* 7: 308–13.

Hirsch, M. and S. Smale (1974) *Differential Equations, Dynamical Systems, and Linear Algebra*. New York: Academic Press.

Hofbauer, J. and K. Sigmund (1988) *The Theory of Evolution and Dynamical Systems*. New York: Cambridge University Press.

Kitcher, P. (1999) "Games Social Animals Play: Commentary on Brian Skyrms' *Evolution on the Social Contract*." *Philosophy and Phenomenological Research* 59: 221–8.

Krebs, D. (2000) "Evolutionary Games and Morality." *Journal of Consciousness Studies* 7: 313–21.

Mar, G. (2000) "Evolutionary Game Theory, Morality and Darwinism." *Journal of Consciousness Studies* 7: 322–6.

Maynard-Smith, J. and G. Price (1973) "The Logic of Animal Conflicts." *Nature* 246: 15–18.

Nesse, R. (2000) "Strategic Subjective Commitment." *Journal of Consciousness Studies* 7: 326–30.

Peixoto, M. M. (1962) "Structural Stability on Two-Dimensional Manifold." *Topology* 1: 101–20.

Proulx, C. (2000) "Distributive Justice and the Nash Bargaining Solution." *Journal of Consciousness Studies* 7: 330–4.

Sacco, P. L. (1995) "Comment." In *The Rational Foundations of Economic Behavior*, ed. K. Arrow et al., 155–71. New York: Macmillan.

Schlag, K. (1998) "Why Imitate, and If So How? A Bounded Rational Approach to the Multi-Armed Bandits." *Journal of Economic Theory* 78: 130–56.

Skyrms, B. (1994) "Darwin meets *The Logic of Decision*." *Philosophy of Science* 61: 503–28.

Skyrms, B. (1996) *Evolution of the Social Contract*. New York: Cambridge University Press.

Skyrms, B. (1997) "Chaos and the Explanatory Significance of Equilibrium: Strange Attractors in Evolutionary Game Dynamics." In *The Dynamics of Norms*, ed. C. Bicchieri et al., 199–222. New York: Cambridge University Press.

Skyrms, B. (1999) "Precis of *Evolution of the Social Contract*" and "Reply to Critics." *Philosophy and Phenomenological Research* 59: 217–20 and 243–54.

Skyrms, B. (2000) "Game Theory, Rationality and Evolution of the Social Contract" and "Reply to Commentary." *Journal of Consciousness Studies* 7: 269–84, 335–9.

Smale, S. (1980) *The Dynamics of Time: Essays on Dynamical Systems, Economic Processes and Related Topics*. New York: Springer.

Taylor, P. and L. Jonker (1978) "Evolutionarily Stable Strategies and Game Dynamics." *Mathematical Biosciences* 40: 145–56.

van Damme, E. (1987) *Stability and Perfection of Nash Equilibria*. Berlin: Springer.

Weibull, J. (1997) *Evolutionary Game Theory*. Cambridge, MA: MIT Press.

5

Dynamics of Conformist Bias

1. Replicator Dynamics

The most thoroughly studied dynamic model for cultural evolution is the replicator dynamics. (Taylor and Jonker 1978. See Hofbauer and Sigmund 1998 for a comprehensive treatment.) It was originally proposed as an account of differential reproduction based on haploid genetics. But it also can be motivated as a model of cultural evolution-based differential imitation (Schlag 1997)—with the more successful strategies being imitated more often. The population is assumed to be large enough so that a deterministic dynamics can be a useful approximation to the true noisy behavior. For strategies A_i we write $U(A_i)$ for the average value to A_i in the population, $P(A_i)$ for the proportion of the population using strategy A_i and $Ubar = \Sigma_i \, P(A_i) \, U(A_i)$ for the average value in the population. Then the replicator dynamics is given by the system of differential equations:

$$dP(A_i)/dt = P(A_i)[U(A_i) - Ubar]$$

Above-average value (fitness, utility) leads to positive growth in population proportion; below-average value leads to a diminishing population proportion.

2. Value

The average value to a strategy may come from average payoffs to that strategy when played against randomly chosen opponents from the population. We will be interested in some well-studied two-person games—The Winding Road, Stag Hunt, Prisoner's Dilemma, Rock-Scissors-Paper. The payoff to one player is determined not only by his own strategy, but also by that of the player with whom he is paired. We assume that the identities of the players are not important, only the strategies played. Thus payoffs are specified by a matrix V_{ij}, which gives the value of A_i when played against A_j. In a large population with random encounters, where all value comes from the game interactions, we can take the average value of a strategy, A_i, to be:

$$\text{Payoff Value} = \Sigma_j P(A_j) A_{ij}$$

This is the analysis familiar from evolutionary game theory.

But in a society with conformist bias, the value of a strategy—that which leads it to be differentially imitated—may have two components: the payoff component and the conformist component. Overall value can be achieved as a weighted average:

$$U(A_i) = (1 - c)\,\text{Payoff Value} + c\,\text{Conformist Value}$$

where the constant c determines the strength of conformist bias. If the conformism constant $c = 1$, conformism counts for everything; if $c = 0$ it counts for nothing.

The conformist value of a strategy increases as the strategy becomes more common. We should have the conformist value of a strategy be some monotonically increasing function of the population proportion of the strategy. Here we will take the simplest choice and let the conformist value simply be the population proportion of the strategy at issue. We can then rewrite overall value as:

$$U(A_i) = (1 - c)\Sigma_j P(A_j) V_{ij} + c\, P(A_i)$$

Overall value then feeds into the replicator dynamics.

For a given type of interaction, we can start with $c = 0$ and then see how the dynamics changes when we add conformist bias. The most dramatic effects that we might encounter consist in the creation of new equilibria or the destruction of old ones. Short of this, we might see changes in the stability of equilibria, with stable equilibria becoming unstable or unstable ones being stabilized. Even if the equilibrium structure is unaltered, the size of basins of attraction for existing equilibria might be changed. Or, perhaps, the addition of conformist bias might leave the dynamical system essentially unchanged.

3. Interactions

The effect of conformist bias will be different relative to different kinds of interactions. Here we survey the evolutionary dynamics of a number of paradigmatic interactions with and without conformist bias.

3.1 Winding Road I

	Right	Left
Right	1	0
Left	0	1

This is a classic pure coordination game that nicely illustrates the virtues of conformism. The choice is to drive on the left or on the right. The conventions of everyone driving on the left or everyone on the right are equally good. It is only important that all choose the same side of the road. Applying the replicator dynamics, we find that the two possible conventions—100 percent left and 100 percent right—are the only stable equilibria, with an additional unstable equilibrium at the point where exactly half the population drives on the left and half on the right. If more than half the population drives on the right, then the dynamics carries it to the equilibrium when all drive on the right; that is to say that the basin of attraction of this equilibrium consists of all population proportions $Pr(R) > .5$. Likewise, the basin of attraction of 100 percent left equilibrium consists of populations with $Pr(L) < .5$.

When we add conformist bias to any degree, nothing changes. The dynamics is exactly the same. The overall value of left is a weighted average of the payoff value of left [= $Pr(L)$] and the conformist value of left [= $Pr(L)$], likewise with right, so the addition of conformist bias adds nothing. The structure of the interaction by itself generates conformism, with or without conformist bias.

3.2 The Stag Hunt

	Stag	Hare
Stag	4	0
Hare	3	3

The Stag Hunt is a coordination game, but not a pure coordination game. There are, as before, three equilibria. The two population monomorphisms, All hunt stag and All hunt hare, are stable attractors in the replicator dynamics. There is also an unstable polymorphic equilibrium at Pr(Stag) = .75. The basin of attraction of the Hare hunting equilibrium [Pr (Stag) < .75] is three times as large as that of the Stag hunting equilibrium [Pr (Stag) > .75]. From the point of view of social welfare this is a shame, because everyone is better off at the Stag Hunting equilibrium.

If we add in some conformist bias, the basic equilibrium structure remains the same—the two stable monomorphisms and the one unstable polymorphism—but the polymorphic equilibrium moves toward the center and the basin of attraction of Hare Hunting is diminished. With 100 percent conformism, the picture would look just like the Winding Road.

Conformist bias has some socially positive effect by decreasing the riskiness of Stag Hunting, and—in a sense—making the dynamical picture more favorable to the socially efficient equilibrium. We can't really generalize from this example, even in coordination games. Consider the following.

3.3 Winding Road II

	Right	Left
Right	3	0
Left	0	1

This is the Winding Road for a population who are all blind in the right eye. There must be a better story, but the point is that although it is still a pure coordination game, one equilibrium is better for everyone than the other. Consequently, the equilibrium where all drive on the right has a greater basin of attraction. If Pr(R) > .25, the dynamics carries the population to All Right; if Pr(R) < .25, the dynamics carries the population to All Left. The dynamical picture looks like that for the Stag Hunt, with Right for Hare and Left for Stag.

If we add conformist bias, the basin of attraction for All Right shrinks and approaches .5 as overall value approaches pure conformism. But here, unlike in the Stag Hunt game, conformism works against mutual benefit rather than for it.

3.4 Rock-Scissors-Paper I

	R	S	P
R	1	2	0
S	0	1	2
P	2	0	1

Rock breaks scissors, scissors cuts paper, paper covers rock. (For a more interesting example of this kind of cyclic structure in a public-goods provision game with optional participation, see Hauert et al. 2002.) The three possible monomorphisms of the population (All Rock, All Scissors, All Paper) do not correspond to Nash equilibria. They are, of course, dynamical equilibria under the replicator dynamics (other types are extinct), but they are dynamically unstable (at each, another type could invade). Thus, at the population state All Rock, a few mutants who play paper could invade.

There is a unique polymorphic equilibrium with 1/3 of the population playing each of the strategies. The first place to look for information about the stability of this equilibrium is the Jacobian matrix of partial derivatives for the dynamics. If all eigenvalues of the Jacobian have negative real part, then the equilibrium is an attractor. If there then is at least one eigenvalue with positive real part, it is unstable. (Hofbauer and Sigmund 1988). The dynamics can be written in terms of $\Pr(R)$ and $\Pr(S)$ since the population proportions must sum to one. Writing the dynamics this way, and evaluating the eigenvalues of the Jacobian at $\Pr(R) = \Pr(S) = 1/3$, we get $[-\mathrm{SQRT}(-1/3), \mathrm{SQRT}(-1/3)]$. The imaginary eigenvalues indicate a rotating motion. Since the real parts of eigenvalues are zero, they do not answer the stability question and other means must be used.

The quantity $\Pr(R)*\Pr(S)*(1-\Pr(R)-\Pr(S))$ is a constant of motion of the system—its time derivative is zero. The orbits of the dynamics must keep this value constant. It assumes its maximum of 1/27 only at the polymorphic equilibrium, $\Pr(R) = \Pr(S) = 1/3$. Off the equilibrium, constant values of the conserved quantity correspond to closed curves around the equilibrium. This is illustrated in the contour plot shown in Figure 5.1. These closed curves are the orbits of the dynamics. The equilibrium is dynamically stable because populations near to it cycle

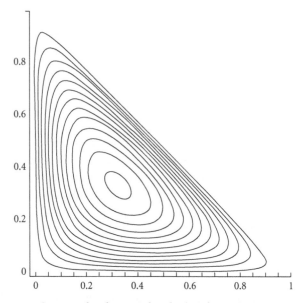

Figure 5.1 Contour plot showing closed orbits for Rock-Scissors-Paper I

around and stay near to it. However, it is not asymptotically stable. Populations near to it are not attracted to it.

If we add even the smallest bit of conformist bias to the dynamics, the polymorphic equilibrium at <1/3,1/3,1/3> is destabilized. The general expression for the eigenvalues of the Jacobian at this point, when conformist bias is included, is:

$$\{c/3 - \text{SQRT}(-1/3 + 2c/3 - c^\wedge 2/3), c/3 + \text{SQRT}(-1/3 + 2c/3 - c^\wedge 2/3)\}$$

If c > 0, then these eigenvalues have positive real part, which indicate that the equilibrium has become unstable. The time derivative of the product of the proportions of the strategies is no longer a constant of motion. Now this quantity decreases along all orbits in the interior of the space of population proportions. If you start arbitrarily near to the equilibrium, the orbit will spiral outward and approach the boundary.

The stability characteristics of the monomorphic equilibria, however, are not changed by a little conformist bias. They remain dynamically unstable saddle points, with Rock, for example, attracting on the edge connecting it with Scissors but repelling along the edge connecting it with Paper.

Adding considerably more conformist bias, however, produces another qualitative change (a bifurcation) in the dynamics. The eigenvalues of the Jacobian at each of the monomorphic equilibria are:

$$\{-c - \text{SQRT}(1 - 2c + c^2), -c - \text{SQRT}(1 - 2c + c^2)\}$$

with no conformist bias, $c = 0$, these are $\{1, -1\}$ indicating the unstable saddle. At $c = .5$, there is a bifurcation, and these values are $\{-1, 0\}$. With $c > .5$, both eigenvalues become negative, indicating that the monomorphisms have changed from (unstable) saddles to (strongly stable) attractors. At the same time, continuity considerations tell us that three new unstable equilibria have been created on the edges. The situation with $c = 0$ and with $c = .6$ are shown in Figures 5.2 and 5.3, with filled circles representing stable equilibria and open circles representing unstable ones.

What have all these dynamical fireworks done for the efficiency of the population? If we measure the results in terms of real payoff, without adding in the supposed satisfaction from conformism, the answer must be "Nothing." The average payoff at the original polymorphic equilibrium <1/3,1/3,1/3> is equal to one. A population at a stable monomorphism still gets the same payoff.

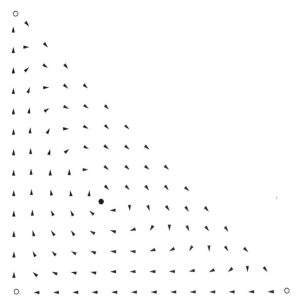

Figure 5.2 No conformist bias (C = 0)

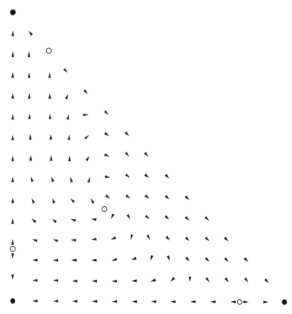

Figure 5.3 Appreciable conformist bias (C = .6)

3.5 Rock-Scissors-Paper II (Zeeman 1980; Hofbauer and Sigmund 1998)

	R	S	P
R	1-e	2	0
S	0	1-e	2
P	2	0	1-e

The original Rock-Scissors-Paper game (without conformist bias) was not structurally stable in the replicator dynamics. So the striking results of conformist bias may have been achieved rather cheaply. In Rock-Scissors-Paper II, with small e, we have a game which is structurally stable, and in which the <1/3,1/3,1/3> polymorphism *is* a stable attractor. Eigenvalues of the Jacobian have negative real part. Furthermore, the product of the population proportions increases along all interior orbits and reaches its maximum at <1/3,1/3,1/3>, so this state is a *global* attractor for all the interior of the space (where no type is extinct). Orbits spiral in to the polymorphic equilibrium. The monomorphisms, All Rock, All Scissors, All Paper, are (unstable) saddles.

As we feed in conformist bias, the dynamics changes qualitatively at $b = e/(1 + e)$. The equilibrium at <1/3,1/3,1/3> ceases to be an attractor, but remains stable. The eigenvalues of the Jacobian now have only imaginary parts. The quantity $Pr(R)^*Pr(S)^*(1 - Pr(R) - Pr(S))$ is again a constant of motion of the system, so that <1/3,1/3,1/3> is surrounded by closed orbits. We are in a situation qualitatively similar to the original Rock-Scissors-Paper without conformist bias.

As conformist bias increases, the system goes through all the changes noted in the discussion under Rock-Scissors-Paper I. At the end, the polymorphism has changed from an attractor to a repellor, the monomorphisms have changed from saddles to attractors, and three new (unstable) equilibria have been created. As before, conformist bias confers no collective benefit. A monomorphic population is no better off (in fact, slightly worse off) than a population at the polymorphic equilibrium.

4. Conclusion

In the simple setting considered in this paper, conformist bias can have dramatic effects on the dynamics of cultural evolution. These effects are sometimes positive, sometimes negative, and sometimes neutral with respect to the collective welfare of the group. Conformist bias is, therefore, not the same as group solidarity. Group solidarity would presumably work for a Pareto-efficient equilibrium—for mutual benefit, but we have seen (most clearly in Section 3.3) that conformist bias can work against it.

The story may, of course, be different when the matter is examined in different settings (compare Boyd and Richerson 1985 and Henrich and Boyd 1998). Interaction between multiple groups can quickly introduce additional complexity. Just moving from one population replicator dynamics to two population replicator dynamics makes a radical difference in Rock-Scissors-Paper games without conformist bias (Sato et al. 2002). Evaluation of the effects of conformist and other types of bias in more complicated settings raises questions well worth pursuing.

References

Boyd, R. and P. J. Richerson (1985) *Culture and the Evolutionary Process.* Chicago, IL: University of Chicago Press.

Hauert, C., S. De Monte, J. Hofbauer, and K. Sigmund (2002) "Volunteering as Red Queen Mechanism for Cooperation in Public Goods Games." *Science* 296: 1129–32.

Henrich, J. and R. Boyd (1998) "The Evolution of Conformist Transmission and the Emergence of Between-Group Differences." *Evolution and Human Behavior* 19: 215–42.

Hofbauer, J. and K. Sigmund (1998) *Evolutionary Games and Population Dynamics.* New York: Cambridge University Press.

Sato, Y., E. Akiyama, and J. D. Farmer (2002) "Chaos in Learning a Simple Two-Person Game." *Proceedings of the National Academy of Sciences* 99: 4748–51.

Schlag, K. H. (1997) "Why Imitate, and If So, How? A Boundedly Rational Approach to Multi-Armed Bandits." *Journal of Economic Theory* 78: 130–56.

Taylor, P. D. and L. Jonker (1978) "Evolutionary Stable Strategies and Game Dynamics." *Mathematical Biosciences* 40: 145–56.

Zeeman, E. C. (1980) "Population Dynamics from Game Theory." In *Global Theory of Dynamical Systems.* Springer Lecture Notes on Mathematics 819. Berlin and New York: Springer.

6

Chaos and the Explanatory Significance of Equilibrium: Strange Attractors in Evolutionary Game Dynamics

1. Introduction

The classical game theory of von Neumann and Morgenstern (1947) is built on the concept of equilibrium. I will begin this chapter with two more or less controversial philosophical claims regarding that equilibrium concept:

1. The explanatory significance of the equilibrium concept depends on the underlying dynamics.
2. When the underlying dynamics is taken seriously, it becomes apparent that equilibrium is not the central explanatory concept.

With regard to the first thesis, let me emphasize a point first made by von Neumann and Morgenstern themselves. Their theory is a static theory which discusses the nature and existence of equilibrium, but which does not address the question: "How is equilibrium reached?" The explanatory significance of the equilibrium concept, however, depends on the plausibility of the underlying dynamics that is supposed to bring players to equilibrium. One sort of story supposes that the decision makers involved reach equilibrium by an idealized reasoning process, which requires a great deal of common knowledge, godlike calculational powers, and perhaps allegiance to the recommendations of a particular theory of strategic interaction. Another kind of story—deriving from evolutionary biology—views game theoretic equilibria as fixed points of evolutionary adaptation, with none of the rational ideal-ization of the first story. The power of game theory to explain a state of affairs as an equilibrium thus depends on the viability of a dynamical scenario appropriate to the situation in question, which shows how such an equilibrium would be reached.

It is well-known that the problem is especially pressing in an area of game theory that von Neumann and Morgenstern did not emphasize: the theory of non-zero sum, non-cooperative games. Here, unlike the zero-sum case, many non-equivalent equilibria are possible. If different deci-sion makers aim for different equilibria, then the joint result of their actions may not be an equilibrium at all. Thus the dynamics must bear the burden of accounting for *equilibrium selection* by the players, because without an account of equilibrium selection the equilibrium concept itself loses its plausibility.

Once one has asked the first dynamical question: "How is equilibrium reached?" it becomes impossible not to ask the more radical question: "Is equilibrium reached?" Perhaps it is not. If not, then it is important to canvass the ways in which may not be reached and explore complex non-convergent behavior permitted by the underlying dynamics. This chapter will take a small step in that direction.

In particular, I will present numerical evidence for extremely compli-cated behavior in the evolutionary game dynamics introduced by Taylor and Jonker (1978). This dynamics, which is based on the process of replication, is found at various levels of chemical and biological organi-zation (Hofbauer and Sigmund 1988). For a taste of what is possible in this dynamics with only four strategies, see the "strange attractor" in Figure 6.1. This is a projection of a single orbit for a four strategy evolutionary game onto the three simplex of the probabilities of the first three strategies. A strange attractor cannot occur in the Taylor–Jonker flow in three strategy evolutionary games because the dynamics takes place on a two dimensional simplex. Zeeman (1980) leaves it open as to whether strange attractors are possible in higher dimensions or not. This paper presents strong numerical evidence for the existence of strange attractors in the lowest dimension in which they could possibly occur.

The plan of the chapter is as follows: Sections 2, 3, and 4 introduce key concepts of games, dynamics, and evolutionary game dynamics. Section 5 will describe the four-strategy evolutionary game that gives

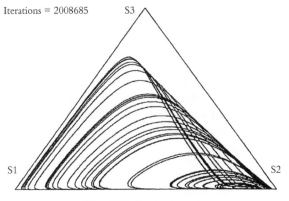

Figure 6.1 Parameter = 5

rise to chaotic dynamics, and the bifurcations that lead to chaos as the parameters of the model are varied. Section 6 will give a stability analysis of the equilibria encountered along the road to chaos described in Section 5. Section 7 describes the numerical calculation of Liapunov exponents. Section 8 indicates some related literature, and discusses the relation to Lotka–Volterra ecological models. My second philosophical claim will be discussed in Section 9.

2. Games

We will be concerned with finite, non-cooperative, normal form games. There are a finite number of players and each player has a finite number of possible strategies. Each player has only one choice to make and makes it without being informed of the choices of any other players. The games are to be thought of as non-cooperative. There is no communication or pre-commitment before the players make their choices. Each possible combination of strategies determines the payoffs for each of the players.

A specification of the number of players, the number of strategies for each player and the payoff function determines the game. A *Nash equilibrium* of the game is a strategy combination such that no player does better on any unilateral deviation. We extend players' possible acts to include randomized choices at specified probabilities over the originally available acts. The new randomized acts are called *mixed strategies*, and the original acts are called *pure strategies*. The payoffs for mixed strategies are defined as their expected values using the probabilities in the mixed acts to define the expectation (and assuming independence between different players acts). We will assume that mixed acts are always available. Then every finite non-cooperative normal form game has a Nash equilibrium.

The game in example 1 has two Nash equilibria in pure strategies, one at <bottom, right> and one at <top, left>. Intuitively, the former equilibrium is—in some sense—highly instable, and the latter equilibrium is the only sensible one.

Example 1

1,1	0,0
0,0	0,0

Selten (1975) introduced the notion of a *perfect equilibrium* to capture this intuition. He considers *perturbed* games wherein each player rather that simply choosing a strategy, chooses to instruct a not perfectly reliable agent as to which strategy to choose. The agent has some small non-zero probabilities for mistakenly choosing each of the strategies alternative to the one he was instructed to choose. Probabilities of mistakes of agents for different players are uncorrelated. An equilibrium in the original game, which is the limit of some sequence of equilibria in perturbed games as the probability of mistakes goes to zero, is called a (trembling-hand) perfect equilibrium. In any perturbed game, see example 1, there is only one equilibrium, with row and column instructing their agents to play top and left and their agents doing so with probability of one minus the small probability of a mistake.

Classical game theory is intended as a theory of strategic interaction between rational human payoff-maximizers. It has sometimes been criticized as incorporating an unrealistically idealized model of human rationality. Maynard-Smith and Price (1973) found a way to apply game theory to model conflicts between animals of the same species. The rationale obviously cannot be that snakes or mule deer are hyperrational, but rather that evolution is a process with a tendency in the direction of increased payoff where payoff is reckoned in terms of evolutionary fitness. A (stable) rest point of such a process must be an optimal point. The insight that just such a tendency may be enough to make rational choice theory and game theory relevant can be carried back to the human realm, and accounts for much of the current interest in dynamic models of learning and deliberation in game theoretic contexts.

Maynard Smith and Price are interested in providing an evolutionary explanation of "limited war" type conflicts between members of the same species, without recourse to group selection. The key notion that they introduce is that of a strategy that would be a stable equilibrium under natural selection, an *evolutionarily stable strategy*. If all members of the population adopt that strategy, then no mutant can invade. Suppose that there is a large population, that contests are pairwise, and that pairing is random. Then the relevant payoff is the average change in evolutionary fitness of an individual, and it is determined by its strategy and the strategy against which it is paired. These numbers can be conveniently presented in a *fitness matrix* and can be thought of as defining the evolutionary game. The fitness matrix is read as giving row's payoff when playing against column.

Example 2

	R	H
R	2	−3
H	−1	−2

Thus, in example 2, the payoff to R when playing against R is 2 but when playing against H is −3. The payoff to H when playing against R is −1 and when playing against H is −2. Here R is an evolutionarily stable strategy because in a population where all members adopt that strategy, a mutant who played H would do worse against members of the population that they would. Likewise, H is an evolutionarily stable strategy, since H does better against H than R does. Suppose, however, that a mutant could do exactly as well against an established strategy as that strategy against itself, but the mutant would do worse against itself than the established strategy. Then the established strategy should still be counted as evolutionarily stable, as it has greater average payoff than the mutant, in a population consisting of players playing it together with a few playing the mutant strategy. This is the formal definition adopted by Maynard-Smith and Price (1972). Let $U(x|y)$ be the payoff to strategy x played against strategy y. Strategy x is *evolutionarily stable* just in case $U(x|x) > U(y|x)$ or $U(x|x) = U(y|x)$ and $U(x|y) > U(y|y)$ for all y different from x. Equivalently, x is evolutionarily stable if:

$$U(x|x) \geq U(y|x)$$
$$\text{If } U(x|x) = U(y|x) \text{ then } U(x|y) > U(y|y)$$

The fitness matrix determines a symmetric payoff matrix for a two-person game—the symmetry deriving from the fact that only the strategies matter, not whether they are played by row or column—as is shown in example 3.

Example 3

2,2	−3,−1
−1,−3	−2,−2

An evolutionarily stable strategy is—by condition 1 above—a symmetric Nash equilibrium of the two-person non-cooperative game. Condition 2 adds a kind of stability requirement.

The formal definition of an evolutionarily stable strategy applies to mixed strategies as well as pure ones, and some fitness matrices will have the consequence that the only evolutionarily stable strategy is a mixed one. This is illustrated in example 4.

Example 4

	H	D
H	−2	2
D	0	1

Neither H nor D is an evolutionarily stable strategy, but a mixed strategy, M, of (1/3) H, (2/3) D is evolutionarily stable. This illustrates condition 2 in the definition of evolutionarily stable strategy. $U(x|M)=2/3$ if x is H or D or any mixture of H and D. But an invader who plays H or D or a different mixture of H and D will do worse against herself that M does against her. For example, consider H as an invader. $U(H|H) = -2$ while

$U(M|H) = -2/3$. The interpretation of mixed strategies as strategies adopted by each member of the population makes sense of the characterization: if all members of the population adopt that strategy, then no mutant can invade. There is an alternative interpretation of mixed strategies in terms of proportions of a polymorphic population, all of whose members play pure strategies. The formal definition of evolutionarily stable strategy in terms of 1 and 2 still makes sense on this reinterpretation of mixed strategies.

If we consider the two-person non-cooperative normal form game associated with a fitness matrix, an evolutionarily stable strategy, x, induces a symmetric Nash equilibrium <x,x> of the game which has certain stability properties. Earlier, we considered Selten's concept of perfect equilibrium, which rules out certain instabilities. Evolutionary stability is a stronger requirement than perfection. If x is an evolutionarily stable strategy, then <x,x> is a perfect symmetric Nash equilibrium of the associated game, but the converse does not hold. In the game associated with the fitness matrix in example 5, <S2,S2> is a perfect equilibrium.[1]

Example 5

	S1	S2	S3
S1	1	0	-9
S2	0	0	-4
S3	-9	-4	-4

S2, however, is not an evolutionarily stable strategy because $U(S1|S2) = U(S2|S2)$ and $U(S1|S1) > U(S2|S1)$.

The concepts of equilibrium and stability in game theory are quasi-dynamical notions. How do they relate to their full dynamical counterparts when game theory is embedded in a dynamical theory of equilibration?

---

[1] But it is not a proper equilibrium. See van Damme (1987) for a definition of proper equilibrium, a proof that if S is an evolutionarily stable strategy, then <S,S> is a perfect and proper equilibrium of the associated game, and a great deal of other information about relations between various stability concepts.

3. Dynamics

The state of a system is characterized by a state vector, $\mathbf{x}$, which specifies the values of relevant variables. (In the case of prime interest here, the relevant variables will be the probabilities of strategies in a game.) The dynamics of the system specifies how the state vector evolves in time. The path that a state vector describes in state space as it evolves according to the dynamics is called a trajectory, or orbit. Time can either be modeled as discrete or as continuous. For the former case, a deterministic dynamics consists of a map which may be specified by a system of difference equations:

$$x(t + 1) = f(x(t))$$

In the latter case, a deterministic dynamics is a flow which may be specified by a system of differential equations:

$$dx/dt = f(x(t))$$

An *equilibrium point* is a fixed point of the dynamics. In the case of discrete time, it is a point, $\mathbf{x}$ of the state space such that $f(\mathbf{x}) = \mathbf{x}$. For continuous time, it is a state, $\mathbf{x} = \langle x_1, \ldots, x_i, \ldots \rangle$ such that $dx_i/dt = 0$, for all i. An equilibrium $\mathbf{x}$ is *stable* if points near to it remain near to it. More precisely, $\mathbf{x}$ is stable if for every neighborhood, V of $\mathbf{x}$, there is a neighborhood, V', of x such that if the state $\mathbf{y}$ is in V' at time $t = 0$, it remains in V for all time $t > 0$. An equilibrium, $\mathbf{x}$, is *strongly stable* (or asymptotically stable) if nearby points tend towards it. That is, to the definition of stability we add the clause that the limit as t goes to infinity of $y(t) = \mathbf{x}$.

An *invariant set* is a set, S, of points of the state space such that if the system starts at a point in S, then at any subsequent time the state of the system is still in S. A unit set is an invariant set just in case its member is a dynamical equilibrium. A closed invariant set, S, is an *attracting set* if nearby points tend towards it; that is, if there is a neighborhood, V, of S such that the orbit of any point in V remains in V and converges to S. An *attractor* is an indecomposable attracting set. (Sometimes other conditions are added to the definition.)

A dynamical system displays *sensitive dependence on initial conditions* at a point if the distance between the orbits of that point and one infinitesimally close to it increases exponentially with time. This sensitivity can be quantified by the *Liapunov exponent*(s) of an orbit. For a one-dimensional map, $x(t+1) = f(x(t))$, this is defined as follows:[2]

$$\lambda = \lim_{n \to \infty} \frac{1}{n} \sum_{i=0}^{n-1} \log_2 \left| \frac{df}{dx} \, at \, x_i \right|$$

A positive Liapunov exponent may be taken as the mark of a *chaotic* orbit. For example, consider the "tent" map:

TENT:

$$x(t+1) = 1 - 2 \left| \frac{1}{2} - x(t) \right|$$

The derivative is defined and its absolute value is 2 at all points except $x = 1/2$. Thus, for almost all orbits the Liapunov exponent is equal to one.

An attractor for which the orbit of almost every point is chaotic is a *strange attractor*. For most known "strange attractors"—like the Lorenz attractor and the Rössler attractor—there is no mathematical proof that they are strange attractors, although the computer experiments strongly suggest that they are. The "strange attractor" in game dynamics which appears in Figure 6.1 and which will be discussed in Sections 5–7 has the same status.

4. Game Dynamics

A number of different dynamical models of equilibration processes have been studied in economics and biology. Perhaps the oldest is the dynamics considered by Cournot (1897) in his studies of oligopoly. There is a series of production quantity setting by the oligopolists, at each time period of which each oligopolist makes her optimal decision on the assumption that the others will do what they did in the last round. The dynamics of the system of oligopolists is thus defined by a *best response map*. A Nash equilibrium is a fixed point of this map. It may be dynamically stable or unstable, depending on the parameters of the Cournot model.

---

[2] For flows the sum is replaced with an integral. For three dimensions, there is a spectrum of three Liapunov exponents, each quantifying divergence of the orbit in a different direction.

A somewhat more conservative adaptive strategy has been suggested by evolutionary game theory. Here we will suppose that there is a large population, all of whose members play pure strategies. The interpretation of a mixed strategy is now as a polymorphism of the population. Asexual reproduction is assumed for simplicity. We assume that individuals are paired at random, that each individual engages in one contest (per unit time), and that the payoff in terms of expected number of offspring to an individual playing strategy S_i against strategy S_j is U_{ij}—given in the ith row and jth column of the fitness matrix, U. The proportion of the population playing strategy S_j will be denoted by $Pr(S_j)$. The expected payoff to strategy i is:

$$U(S_i) \sum_j pr(S_j) U_{ij}$$

The average fitness of the population is:

$$U(Status\, Quo) = \sum_i pr(S_i) U(S_i)$$

The interpretation of payoff in terms of Darwinian fitness then gives us a map for the dynamics of evolutionary games in discrete time:

$$pr'(S_i) = pr(S_i) \frac{U(S_i)}{U(Status\ Quo)}$$

(where pr' is the proportion in the next time period).

The corresponding flow is given by:

$$\frac{pr'(S_i)}{dt} = pr(S_i) \frac{U(S_i) - (Status\ Quo)}{U(Status\ Quo)}$$

As long as we are concerned—as we are here—only with symmetric evolutionary games, the same orbits are given by a simpler differential equation:

$$\frac{pr'(S_i)}{dt} = pr(S_i)[U(S_i) - U(StatusQuo)]$$

This equation was introduced by Taylor and Jonker (1978) to provide a dynamical foundation for the quasi-dynamical notion of evolutionarily stable strategy of Maynard Smith and Price (1973). It has subsequently been studied by Zeeman (1980), Hofbauer (1981), Bomze (1986), van Damme (1987), Hofbauer and Sigmund (1988), Samuelson (1988), Crawford (1989), and Nachbar (1990). It will be the dynamics considered in our example in the next section. It worth noting that even though the Taylor–Jonker dynamics is motivated by context where the payoffs are measured on an absolute scale of evolutionary fitness, nevertheless the orbits in phase space (although not the velocity along these orbits) is invariant under a linear transformation of the payoffs. Thus the Taylor–Jonker dynamics may be of some interest in contexts for which it was not intended, where the payoffs are given in von Neumann–Morgenstern utilities.

Relying on the foregoing studies, I will briefly summarize some of the known relations between quasi-dynamical equilibrium concepts and dynamical equilibrium concepts for this dynamics. If [M,M] is a Nash equilibrium of the two-person non-cooperative game associated with an evolutionary game, then M is a dynamic equilibrium of the Taylor–Jonker flow. The converse is not true, since every pure strategy is an equilibrium of the flow. However, if an orbit starts at a completely mixed point and converges to a pure strategy then that strategy is a Nash equilibrium. Furthermore, if M is a stable dynamic equilibrium in the Taylor–Jonker flow, then [M,M] must be a Nash equilibrium of the associated game. However if M is dynamically stable, [M,M] need not be perfect, and if [M,M] is perfect, then M need not be dynamically stable. If M is dynamically strongly stable (asymptotically stable) then [M,M] must be perfect, but the converse does not hold. If M is an evolutionarily stable strategy in the sense of Maynard-Smith and Price then it is perfect, but the converse does not hold. We do have equivalence between an evolutionarily stable strategy and a strongly dynamically stable strategy in the special case of two strategy evolutionary games, but already in the case of three strategies there can be a strongly dynamically stable polymorphic population that is not a mixed evolutionarily stable strategy. Thus, although there are important relations here between the quasi-dynamical and dynamical equilibrium concepts, they tend to draw the line at somewhat different places.

As an example of a third kind of dynamics, we mention the fictitious play of Brown (1951). Like the Cournot dynamics, there is a process in discrete time, at each stage of which each player plays a strategy which maximizes expected utility, according to her beliefs. But these beliefs are not quite so naive as those of the Cournot player. Rather than proceeding on the assumption that all other players will do just what they did last time, Brown's players form their probabilities of another player's next act according to the proportion of times that player has played that strategy in the past.[3] Brown interpreted his as fictitious play, and Cournot interpreted his as real play, but either could just as well be interpreted the other way. Thorlund-Peterson (1990) studies a dynamics closely related to Brown's in the context of a Cournot oligopoly, where it is shown to have convergence properties superior to those of the Cournot dynamics. Brown's dynamics is driven by a simple inductive rule: Use the observed relative frequency as your probability. The basic scheme could be implemented using modified inductive rules. A class of simple Bayesian inductive rules which share the same asymptotic properties as Brown's rule are investigated in Skyrms (1991). For these models, if the dynamics converges it converges to a Nash equilibrium in undominated strategies. For two-person games, such an equilibrium must be *perfect*. This contrasts with the Taylor–Jonker dynamics where an orbit can converge to a dynamically stable equilibrium, M, where [M,M] is an imperfect equilibrium of the corresponding two-person non-cooperative game.

5. The Road to Chaos

In this section we will focus on the Taylor–Jonker flow. Flows are usually better behaved than the corresponding maps, but we will see that this dynamics is capable of quite complicated behavior. Taylor and Jonker already note the possibility of non-convergence because of oscillations in three-strategy evolutionary games. They consider the game whose fitness matrix, U, is given in example 6 (where a is a parameter to be varied):

Example 6

	S1	S2	S3
S1	2	1	5
S2	5	a	0
S3	1	4	3

For a = 1 the completely mixed equilibrium serves as an example of an equilibrium which is dynamically strongly stable but not an

---

[3] To make the dynamics autonomous, expand the concept of state of the system to include a "memory" of frequencies of past plays.

evolutionarily stable strategy. For a < 3 the equilibrium is strongly stable, but at a = 3 a qualitative change takes place. Now the mixed equilibrium is stable but not strongly stable. It is surrounded by closed orbits. At a > 3 the mixed equilibrium is unstable and the trajectories spiral outward to the boundary of the space. The change that takes place at a = 3 is a *degenerate* Hopf bifurcation. (See Guckenheimer and Holmes 1986: 73, 150 ff.) It is degenerate because the situation at a = 3 is not structurally stable. Any small perturbation of the value of a destroys the closed orbits. This is just about as wild as the dynamical behavior can get with three strategies. In particular, *generic* Hopf bifurcations are impossible here. (See Zeeman 1980; Hofbauer 1981. Zeeman proves that a generic Hopf bifurcation is impossible for three-strategy games, and describes the structurally stable flows for such games under the assumption that is discharged in Hofbauer.) And chaotic strange attractors are not possible, because the flow takes place on a two dimensional simplex.

However, with four strategies we get the strange attractor pictured in Figure 6.1. (This is a projection of the three dimensional simplex of probabilities for four strategies onto the two dimensional simplex of the first three strategies. The three dimensional structure, however, is fairly easy to see in the figure.) There is a route to this strange attractor that leads through a generic Hopf bifurcation. Consider the fitness matrix, U, of example 7 (where a is the parameter to be varied):

Example 7

−1	−1	−10	1,000
−1.5	−1	−1	1,000
a	.5	0	−1,000
0	0	0	0

Figures 6.1 through 6.6 are snapshots taken along the path to chaos as this parameter is varied. At a = 2.4 there is convergence to a mixed equilibrium as shown in Figure 6.2. The orbit spirals in towards the mixed equilibrium, which is visible as the white dot in the center of the orbit. As the value of a is raised there is a generic Hopf bifurcation giving rise to a limit cycle around the mixed equilibrium. This closed orbit is structurally stable; it persists for small variations in the parameter. It is also an attracting set. This closed orbit is shown for a = 2.55 in Figure 6.3.

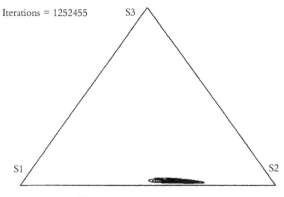

Figure 6.2 Parameter = 2.4

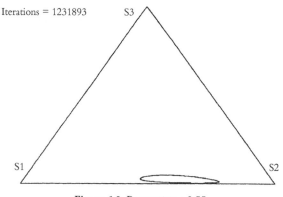

Figure 6.3 Parameter = 2.55

As the value of the parameter is raised further, the limit cycle expands and then undergoes a period doubling bifurcation. Figure 6.4 shows the cycle of period 2 at a = 3.885. This is followed by another period doubling bifurcation, leading to a cycle of period 4 at a = 4.0, as shown in Figure 6.5. There are very long transients before the orbit settles down to this cycle. At a = 5, we get a transition to chaotic dynamics on the strange attractor shown in Figure 6.1. Raising the parameter to a = 6 leads to further geometrical complications in the strange attractor as shown in Figure 6.6.

Differential equations were numerically integrated in double precision using fourth-order Runge-Kutta method (Press et al, 1989). For Figures 6.1

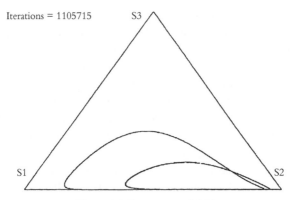

Figure 6.4 Parameter = 3.885

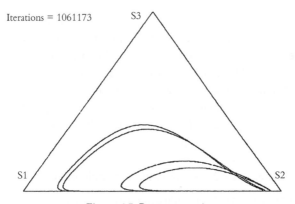

Figure 6.5 Parameter = 4

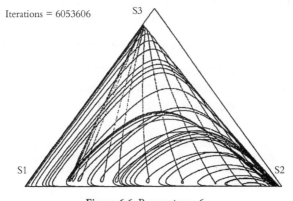

Figure 6.6 Parameter = 6

through 6.4 and 6.6 a fixed step size of .001 was used. For Figure 6.5 a fixed step size of .01 was used. This was done on an IBM model 70 personal computer with a 387 math coprocessor. The projection of the orbit on the simplex of probabilities of the first three strategies was plotted to the screen in vga graphics mode. For Figures 6.1 through 6.4, the first 50,000 steps (= 50 time units) were not plotted to eliminate transients. For Figure 6.5, the first 100,000 (=1,000 time units) steps were omitted to eliminate very long transients. For Figure 6.6, only the first 1,000 steps were omitted. In each case, the total number of steps run is shown in the top left corner of the illustration. The screen was captured using the WordPerfect 5.1 GRAB utility and printed on a Hewlett Packard LaserJet II.

6. Stability Analysis of Equilibria

As a supplement and a check on the graphical information presented in the previous section, the interior equilibrium points along the route to chaos were calculated in high precision (40 decimal places) using *Mathematica*. The Jacobian matrix of partial derivatives was then evaluated at the equilibrium point, and its eigenvalues found.

These are used in stability analysis of the equilibria. (See Hirsch and Smale 1974: ch. 6.) One of these eigenvalues will always be zero; it is an artifact of the constraint that probabilities add to one, and is irrelevant to the stability analysis.[4]

For example, at $a = 2$ there is an interior equilibrium at:

$x1 = 0.5136399954343111516950119849332268000594$
$x2 = 0.4565688848304988015066773199406460044972$
$x3 = 0.0285355553019061750941673324962903778108$
$x4 = 0.0012555644332838717041433626298367766236$

and at this point the eigenvalues of the Jacobian found numerically to be:

$-0.8576108025804070626366665715951399308,$
$-0.0562990388422014452825117944612367686+$
$0.2875123366474160989192752729129540\,\mathrm{I}$

---

[4] See Bomze (1986: 48) or van Damme (1987: 222) and note that in the example given, the expected utility of the status quo (= the average population fitness) at a completely mixed equilibrium point must be equal to zero, since for this fitness matrix, the expected utility of strategy 4 is identically zero.

−0.0562990388422014452825117944612367686−
0.28751233664741609891927527291295404 I
−5.42045724163219646523489178011128836 * 10^{-42}

The last eigenvalue is the insignificant zero eigenvalue. The significant eigenvalues all have negative real parts, indicating a strongly stable equilibrium, which attracts much in the way illustrated in Figure 6.2. Indeed at a = 2.4—the situation actually illustrated in Figure 6.2—the situation is qualitatively much the same. The equilibrium has moved to about:

x1 = 0.363942
x2 = 0.614658
x3 = 0.020219
x4 = 0.001181

(henceforth I suppress the full precision in reporting the results). The non-significant zero eigenvalue of the Jacobian is numerically calculated at the order of 10^{-39}. The significant eigenvalues are approximately:

−0.9752593,
−0.001670447 + 0.26020784 I
−0.001670447 − 0.26020784 I

However, when we move to the limit cycle illustrated in Figure 6.3 at a = 2.55, the situation changes drastically. The equilibrium is now at approximately:

x1 = 0.328467
x2 = 0.653285
x3 = 0.018248
x4 = 0.001164

and the significant eigenvalues of the Jacobian are:

−.993192,
0.00572715 + 0.250703 I,
0.00572715 − 0.250703 I

The real eigenvalue is negative but the imaginary eigenvalues have positive real parts. Thus the equilibrium is an unstable saddle, with the imaginary eigenvalues indicating the outward spiral leading to the limit cycle. A little trial and error in this sort of computation indicates that the

Hopf bifurcation, where the real parts of the imaginary eigenvalues pass from negative to positive, takes place between a = 2.41 and a = 2.42, where the real parts of the imaginary eigenvalues are respectively about −0.001 and +0.001.

In the chaotic situation where a = 5 shown in Figure 6.1, the equilibrium has now moved to approximately:

x1 = 0.12574
x2 = 0.866212
x3 = 0.006956
x4 = 0.001070

The eigenvalues of the Jacobian are:

−1.0267,
0.173705 + 0.166908 I
0.173705 + 0.166908 I

This still indicates a saddle point equilibrium but here—as shown in Figure 6.1—the orbit passes very close to this unstable equilibrium point.

7. Numerical Calculation of Liapunov Exponents

Liapunov exponents were calculated numerically using the algorithm presented in Wolf et al. (1985: Appendix A). This integrates the differential equations of the dynamical system to obtain a fiducial trajectory, and simultaneously integrates four copies of the linearized differential equations of the system with coefficients determined by the location on the fiducial trajectory, to calculate the Liapunov spectrum. The latter are started at points representing a set of orthonormal vectors in the tangent space, and are periodically reorthonormalized during the process. In the calculation, logarithms are taken to the base 2. The code was implemented for the replicator dynamics by Linda Palmer. Differential equations were integrated in double precision using the IMSL Library integrator DIVPRK. The program was tested running it at a = 2, starting it on the attracting equilibrium. In this case, the spectrum of Liapunov exponents (when converted to natural logarithms) should just consist of the real parts of the eigenvalues of the Jacobian evaluated at the equilibrium, which were discussed in the last section. The experimental results of a run from t = 0 to t = 110,000 were in agreement with the theoretical results up to four or five decimal places:

Experimental Results	Theoretical Results
−0.85761	−0.85761
−0.0563	−0.0563
−0.0563	−0.0563
$-3.4 * 10^{-6}$	0

The three negative exponents indicate the attracting nature of the equilibrium point, and the zero exponent corresponds to the spurious eigenvalue as explained in the last section.

For a limit cycle in three dimensions, the Liapunov spectrum should have the qualitative character <0, −, −>. The experimental results on the limit cycles at a = 2.55, a = 3.885 and a = 4 have the appropriate qualitative character. Dropping one spurious zero exponent, we are left with:

	a = 2.55	a = 3.885	a = 4
L1	0.000	0.000	0.000
L2	−0.020	−0.008	−0.004
L3	−1.395	−1.419	−1.423

For a strange attractor in three dimensions, the Liapunov exponents should have the qualitative character <+, 0, −>. At a = 5, where visually we see the onset of chaos in Figure 6.1, the Liapunov spectrum was calculated on a number of runs on a number of computers varying the reorthonormalization frequency and various parameters of the differential equation integrator. Dropping one spurious zero exponent, the following results are very robust:

L1: 0.010
L2: 0.000
L3: −1.44

For a "gold standard run" the equations were integrated from t = 0 to t = 1,000,000 with an error tolerance of 10^{-11}. On this run the zeros (both L2 and the spurious exponent) are zeros to six decimal places. Details of the convergence are shown graphically in Figures 6.7–6.10

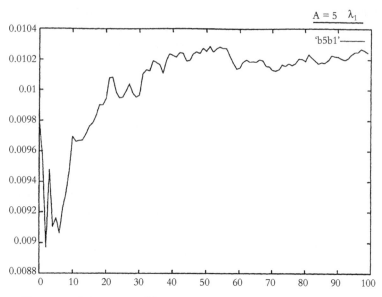

Figure 6.7 Convergence of first Liapunov exponent to a positive quantity

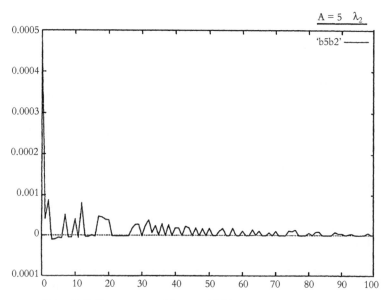

Figure 6.8 Convergence of the second Liapunov exponent to zero

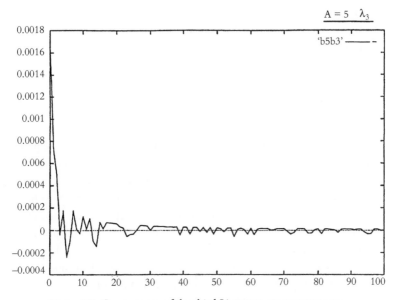

Figure 6.9 Convergence of the third Liapunov exponent to zero

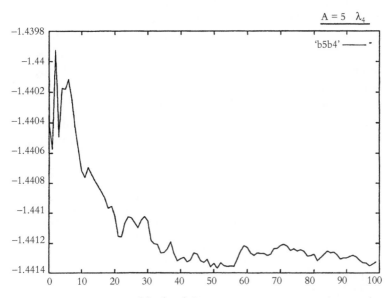

Figure 6.10 Convergence of the fourth Liapunov exponent to a negative quantity

(where one unit on the x axis represents 10,000 units of time). The positive value of the largest Liapunov exponent, L1, indicates that there has indeed been a transition to chaos.[5]

8. Relation to Lotka–Volterra Models and to Other Literature

There are two papers discussing chaos in different dynamics for equilibration in games: one in an economic context and one in the context of theoretical computer science. Rand (1978) considers Cournot duopoly models where the dynamics is Cournot's best response map. Where the reaction functions are tent shaped and have slope greater than one, we get chaotic dynamics. This model differs from the one in the previous section in a number of ways: (1) It is a map rather than a flow that is considered; (2) it is a different dynamics; (3) when the Cournot model is considered as a game, there are an infinite number of pure strategies. Huberman and Hogg (1988) are concerned with distributed processing on computer networks. The problem of efficient use of resources in a network is modeled as a finite game, and a quasi-evolutionary account of the dynamics of adaptation is proposed. In particular, they argue for chaos in the limit of long delays for a delay differential equation modeling information lag. The argument is that the long-term behavior is modeled by a difference equation which is in a class all of whose members display chaotic behavior. The setting considered by Huberman and Hogg is conceptually closer to the one in this chapter than that of Rand in that only finite games are considered, but the dynamics is different.

There is a closer connection with ecological models which do not, on the face of them, have much to do with evolutionary game theory. These

---

[5] For purposes of comparison, the largest Liapunov exponent here is roughly an order of magnitude smaller than that of the Rössler attractor. But the mean orbital period of the attractor is roughly an order of magnitude larger. If we measured time in terms of mean orbital periods, L1 would here be of the same order of magnitude as L1 for the Rössler attractor. Data on the Rössler attractor were obtained from Wolf et al. (1985).

At $a = 6$, although the attractor appears to become geometrically more complex, the Liapunov spectrum is little changed:

L1: 0.009
L2: 0.000
L3: −1.44

are the Lotka–Volterra differential equations which are intended as simple models of population interactions between different species. For n species, they are:

$$\frac{dx_i}{dt} = x_i \left[r_i + \sum_{j=1}^{n} a_{ij}x_j \right]$$

The x_i are the population densities, the r_i the intrinsic growth or decay rates for a species, and the a_{ij} the interaction coefficients which give the effect of the jth species on the ith species.

The dynamics of two species Lotka–Volterra systems—either predator prey or two competing species—is well understood, and the dynamics of three and higher dimensional Lotka–Volterra systems is a subject of current research. Unstable cycles are possible in two dimensional (predator-prey) Lotka–Volterra systems, but chaos is not. In three dimensions, however, several apparent strange attractors have been found. The first was found by Vance (1978) and classified as spiral chaos by Gilpin (1979). "Gilpin's strange attractor" has been extensively studied by Shaffer (1985), Shaffer and Kot (1986), and Vandermeer (1991). Other strange attractors have been reported in three dimensional Lotka–Volterra systems. Arneodo et al. (1980, 1982) use a mixture of numerical evidence and theoretical argument to support the hypothesis of Silnikov-type strange attractors in three and higher dimensions. See also Takeuchi and Adachi (1984) and Gardini et al. (1989). May and Leonard (1975) show that other kinds of wild behavior are possible in Lotka–Volterra systems of three competitors. Smale (1976) shows that for ecological systems modeled by a general class of differential equations (not necessarily Lotka–Volterra) any kind of asymptotic dynamical behavior—including the existence of strange attractors—is possible if there are five or more competing species.

There is an intimate connection between the Taylor–Jonker game dynamics and the Lotka–Volterra dynamics, which is established by Hofbauer (1981). A Lotka–Volterra system with n species corresponds to an evolutionary game with n+1 strategies such that the game dynamics on the evolutionary game is a topologically orbital equivalent to the Lotka–Volterra dynamics. To each species in the Lotka–Volterra system, there is a ratio of probabilities of strategies in the game with the same dynamics. Thus it is possible to use known facts about one kind of dynamical system to establish facts about the other. Hofbauer uses the known fact that two-species Lotka–Volterra systems do not admit limit cycles to verify Zeeman's conjecture that three-strategy evolutionary games do not admit stable limit cycles in the game dynamics. It is thus possible to investigate game dynamical pathology with an eye towards ecological pathology. The strange attractor of the previous section is, in fact, the game theoretic counterpart to Gilpin's strange attractor. For a game dynamical counterpart of the attractor of Arneodo et al. (1980) we have example 8:

Example 8			
0	−.6	0	1
1	0	0	−.5
−1.05	−.2	0	1.75
.5	−.1	.1	0

9. Conclusion

Let us return to the second philosophical thesis with which I began this chapter. That is that *When the underlying dynamics is taken seriously, it becomes apparent that equilibrium is not the central explanatory concept.* Rather, I would take the central dynamical explanatory concept to be that of an *attractor* (or attracting set). Not all dynamical equilibria are attractors. Some are unstable fixed points of the dynamics. In the dynamical system of example 7 with a = 5 there is an unstable equilibrium point which is never seen. And not all attractors are equilibria. There are limit cycles, quasi-periodic attractors, and strange attractors. The latter combine a kind of internal instability with macroscopic asymptotic stability. Thus, they can play the same kind of explanatory role as that of an attracting equilibrium—although what is explained is a different kind of phenomenon.

Even this latter point, however, must be taken with a grain of salt. That is because of the possibility of extremely long transients. In example 7 with a = 4, if we had omitted only the first 50 time units, we would not have eliminated the transient, and the plot would have looked like the strange attractor of Figure 6.1 rather than one of a limit cycle. If transients are long enough, they may govern the phenomenae of interest to us. The concept of an attractor lives at t = infinity, but we do not.

Acknowledgments

The existence of this strange attractor together with a preliminary study of the route to chaos involved was first reported in Skyrms (1992). This paper contains further experimental results. I would like to thank the University of California at Irvine for support in the form of computing time and Linda Palmer for implementing and running programs to determine the Liapunov spectrum. I would also like to thank Immanuel Bomze, Vincent Crawford, William Harper, and Richard Jeffrey for comments on an earlier version of this paper.

References

Arneodo, A., P. Coullet, and C. Tresser (1980) "Occurrence of Strange Attractors in Three Dimensional Volterra Equations." *Physics Letters* 79: 259–63.

Arneodo, A., P. Coullet, J. Peyraud, and C. Tresser (1982) "Strange Attractors in Volterra Equations for Species in Competition." *Journal of Mathematical Biology* 14: 153–7.

Bomze, I. M. (1986) "Non-cooperative 2-person Games in Biology: a Classification." *International Journal of Game Theory* 15: 31–59.

Brown, G. W. (1951) "Iterative Solutions of Games by Fictitious Play." In *Activity Analysis of Production and Allocation* (Cowles Commission Monograph), 374–76. New York: Wiley.

Cournot, A. (1897) *Researches into the Mathematical Principles of the Theory of Wealth* (tr. from the French ed. of 1838). New York: Macmillan.

Crawford, V. (1989) "Learning and Mixed-Strategy Equilibria in Evolutionary Games." *Journal of Theoretical Biology* 140: 537–50.

van Damme, E. (1987) *Stability and Perfection of Nash Equilibria*. Berlin: Springer.

Gardini, L., R. Lupini, and M. G. Messia (1989) "Hopf Bifurcation and Transition to Chaos in Lotka–Volterra equation." *Mathematical Biology* 27: 259–72.

Gilpin, M. E. (1979) "Spiral Chaos in a Predator-Prey Model." *The American Naturalist* 13: 306–8.

Guckenheimer, J. and P. Holmes (1986) *Nonlinear Oscillations, Dynamical Systems and Bifurcations of Vector Fields* (Corrected second printing). Berlin: Springer.

Hirsch, M. W. and S. Smale (1974) *Differential Equations, Dynamical Systems and Linear Algebra*. New York: Academic Press.

Hofbauer, J. (1981) "On the Occurrence of Limit Cycles in the Volterra–Lotka Equation." *Nonlinear Analysis* 5: 1003–7.

Hofbauer, J. and K. Sigmund (1988) *The Theory of Evolution and Dynamical Systems*. Cambridge: Cambridge University Press.

Huberman, B. A. and T. Hogg (1988) "Behavior of Comptutational Ecologies." In *The Ecology of Computation*, ed. B. A. Huberman, 77–115. Amsterdam: North Holland.

May, R. M. and W. L. Leonard (1975) "Nonlinear Aspects of Competition between Three Species." *SIAM Journal of Applied Mathematics* 29: 243–53.

Maynard Smith, J. (1982) *Evolution and the Theory of Games*. Cambridge: Cambridge University Press.

Maynard Smith, J. and G. R. Price (1973) "The Logic of Animal Conflict." *Nature* 146: 15–18.

Nachbar, J. H. (1990) " 'Evolutionary' Selection Dynamics in Games: Convergence and Limit Properties." *International Journal of Game Theory* 19: 59–89.

Press, J., B. Flannery, S. Teukolsky, and W. Vetterling (1989) *Numerical Recipes: The Art of Scientific Computing*, rev. ed. Cambridge: Cambridge University Press.

Rand, D. (1978) "Exotic Phenomena in Games and Duopoly Models." *Journal of Mathematical Economics* 5: 173–84.

Rössler, O. (1976) "Different Types of Chaos in Two Simple Differential Equations." *Zeitschrift fur Naturforschung* 31a: 1664–70.

Samuelson, L. (1988) "Evolutionary Foundations of Solution Concepts for Finite, Two-Player, Normal-Form Games." In *Theoretical Aspects of Reasoning About Knowledge*, ed. M. Vardi. San Mateo, CA: Morgan Kaufmann.

Selten, R. (1975) "Reexamination of the Perfectness Concept of Equilibrium in Extensive Games." *International Journal of Game Theory* 4: 25–55.

Shaffer, W. M. (1985) "Order and Chaos in Ecological Systems." *Ecology* 66: 93–106.

Shaffer, W. M. and M. Kot (1986), "Differential Systems in Ecology and Epidemiology." In *Chaos: An Introduction* ed. A. V. Holden, 158–78. Manchester: University of Manchester Press.

Skyrms, B. (1992) "Chaos in Game Dynamics." *Journal of Logic, Language and Information* 1: 111–30.

Smale, S. (1976), "On the Differential Equations of Species in Competition." *Journal of Mathematical Biology* 3: 5–7.

Takeuchi, Y. and N. Adachi (1984) "Influence of Predation on Species Coexistence in Volterra Models." *Mathematical Biosciences* 70: 65–90.

Taylor, P. and L. Jonker (1978) "Evolutionarily Stable Strategies and Game Dynamics." *Mathematical Biosciences* 40: 145–56.

Thorlund-Peterson, L. (1990) "Iterative computation of Cournot equilibrium." *Games and Economic Behavior* 2: 61–75.

Vance, R. R. (1978) "Predation and Resource Partitioning in a one-predator-two prey model community." *American Naturalist* 112: 441–8.

Vandermeer, J. (1991) "Contributions to the Global Analysis of 3-D Lotka–Volterra Equations: Dynamic Boundedness and Indirect Interactions in the Case of One Predator and Two Prey." *Journal of Theoretical Biology* 148: 545–61.

von Neumann, J. and O. Morgenstern (1947) *Theory of Games and Economic Behavior*. Princeton: Princeton University Press.

Wolf, A., J. B. Swift, H. L. Swinney, and J. A. Vastano (1985) "Determining Lyaponov Exponents from a Time Series." *Physica* 16-D: 285–317.

Zeeman, E. C. (1980) "Population Dynamics from Game Theory." In *Global Theory of Dynamical Systems*, ed. Z. Niteck and C. Robinson (Lecture Notes in Mathematics 819), 471–97. Berlin: Springer.

7

Evolutionary Dynamics of Collective Action in N-person Stag Hunt Dilemmas

with Jorge M. Pacheco, Francisco C. Santos, and Max O. Souza

1. Introduction

During recent years, evolutionary game theory has been able to provide key insights into the emergence and sustainability of cooperation at different levels of organization (Axelrod and Hamilton 1981; Maynard-Smith 1982; Axelrod 1984; Boyd and Richerson 1985; Hofbauer and Sigmund 1998; Skyrms 2001, 2004; Macy and Flache 2002; Hammerstein 2003; Nowak and Sigmund 2004; Nowak et al. 2004; Santos and Pacheco 2005; Nowak 2006; Ohtsuki et al. 2006; Santos et al. 2006). The most popular and studied game has been the two-person Prisoner's Dilemma (PD). However, other social dilemmas, such as the snowdrift game (Sugden 1986) or the Stag Hunt (SH) game (Skyrms 2004) also constitute powerful metaphors for many situations routinely encountered in the natural and social sciences (Macy and Flache 2002; Skyrms 2004).

In particular, the SH game constitutes the prototypical example of the social contract, and one can identify instances of SH games in the writings of, for example, Rousseau, Hobbes, and Hume (Skyrms 2004). Maynard-Smith and Szathmáry (1995) have discussed the social contracts implicit in some of the major transitions of evolution. After framing most of the discussion in terms of the PD, they remarked that perhaps the SH (their rowing game) is a better model. In an SH there is an equilibrium in which both players cooperate as well as one in which both defect.

Whenever collective action of groups of individuals is at stake, *N*-person games are appropriate. Recent literature has focused on *N*-person Prisoner's Dilemmas (NPDs) in the form of provision of public-goods games (PGGs) (Kollock 1998; Hauert et al. 2002, 2006, 2007; Brandt et al. 2006; Milinski et al. 2006, 2008; Rockenbach and Milinski 2006; Santos et al. 2008). The prototypical example of a PGG is captured by the so-called NPD. It involves a group of *N* individuals, who can be either cooperators (C) or defectors (D). Cs contribute a cost *c* to the public good, whereas Ds refuse to do so. After all individuals are given the chance to contribute, the accumulated contribution is multiplied by an enhancement factor *F*, and the total amount is equally shared among all individuals of the group. In other words, if there were *k* Cs in a group of *N* individuals, Ds end up with kFc/N, whereas Cs only get $kFc/N - c$, that is, in mixed groups Cs are always worse off than Ds. If *F* is smaller than *N*, to cooperate is always disadvantageous against any combination of actions by other group members. In this sense, we have an NPD. Evolutionary game theory directly leads to the tragic outcome in which everybody ends up defecting, hence foregoing the public good. When the group is a mere pair of individuals, this dilemma reduces to the two-person PD.

Consider, however, group hunts of three or four lionesses in Etosha National Park, Namibia (Stander 1992). Two lionesses, the *wings*, attack a group of prey from either side panicking them to run forward. They run right into one or two other lionesses, positioned as *centers*, who are waiting for them. This kind of hunt is highly successful. It is not possible with one or two participants, but it is with three and is even better with four. This is not a generalized PD, but a generalized SH. It is an SH because, unlike the PD, there is a cooperative equilibrium where if others do their part, it is best for you to do yours as well.

Variations on this kind of cooperative hunting have been observed in other species, such as chimpanzees in the Tai forest (Boesch 2002) and African wild dogs (Creel and Creel 1995). In animals, other collective actions, such as lions defending a kill against a pack of hyenas, can also be seen as generalized SH games (Maynard-Smith and Szathmáry 1995).

In human affairs, we also find collective action problems that can be viewed as generalized SHs, not only in literal hunts such as the whale hunts discussed in Beding (2008), but also in international relations (Jervis 1978) and macroeconomics (Bryant 1994).

Back to the lionesses in Etosha National Park, two individuals are not enough for a cooperative hunt, three can be successful and four even more so. The average payoff of an individual depends on the number of participants and may vary according to species and environment. Much empirical evidence supports a U-shaped function for average meat per participant across a number of species, but it is controversial whether this remains true when energetic costs of the hunt are taken into account (Creel and Creel 1995; Packer and Caro 1997).

Here we focus on games where there is a threshold (M) for participants below which no public good is produced. We do not make the general assumption that total participation gives each individual the highest payoff. For instance, we include the possibility of "three in a boat, two must row" (Taylor and Ward 1982; Ward 1990), a generalization of the SH game to three players, where contributions of two out of three players are required for the success of the joint venture. If two others row, there is an incentive to free ride; but if one other rows, there is an incentive to jump in and contribute. There may be an analog in cooperative hunting by lions in richer environments where prides are larger and the participation of the entire group is not so helpful.

We shall start by investigating the evolutionary dynamics of Cs and Ds in the traditional setting of evolutionary game theory, that is, infinite well-mixed populations evolving. The fitness of individuals is determined by their payoff collected when engaging in an N-person Stag Hunt (NSH) dilemma requiring at least $M < N$ individuals to produce any public good at all. We shall find that the NSH game leads to richer and more interesting evolutionary dynamics scenarios than the corresponding NPD. Subsequently, we investigate the implications of taking into account the fact that populations are finite. Evolutionary dynamics for large finite populations was pioneered in economics by Young (1993) and by Kandori et al. (1993). The focus here is on the limiting effect of mutation as it becomes infrequent. Owing to mutation, evolutionary dynamics becomes an ergodic Markov chain (Nowak et al. 2004). In the classic SH, it is shown that the population spends almost all its time at the non-cooperative equilibrium.

Evolutionary dynamics of a growing (or shrinking) finite population with random deaths is modeled in Schreiber (2001) and by Benaim et al.

(2004). Either a strategy or the whole population can wander into extinction, but if this does not happen the trajectory of the growing population comes to approximate that of the replicator dynamics.

We shall focus on a (possibly small) well-mixed population of fixed size Z without mutation. The dynamics will be a Markov process, with the only possible end states—the absorbing states—being monomorphisms. When the population is large, the dynamics approximates the replicator dynamics in the medium run, but it will eventually end up in one of the absorbing states. Thus, it may spend a long time near a stable polymorphic equilibrium of the associated mean-field dynamics before eventually being absorbed by a monomorphism. For small populations where population size is close to group size, there is also the "spite" effect first noted by Hamilton (1970), which works against cooperation.

2. Results

(a) Evolutionary dynamics in infinite populations

Let us assume an infinite, well-mixed population, a fraction x which is composed of Cs, the remaining fraction $(1 - x)$ being Ds, and let us further assume that the groups of N individuals are sampled randomly from the population. As shown in Appendix A, random sampling of individuals leads to groups whose composition follows a binomial distribution (Hauert et al. 2006), which also establishes the average fitness of Cs (f_C) and Ds (f_D). In each N-individual group with k Cs, the fitness of Ds is given by $\Pi_D(k) = (kFc/N)\theta(k - M)$, where the Heaviside step function ($\theta(x)$ satisfies $\theta(x < 0) = 0$ and $\theta(x \geq 0) = 1$. The corresponding fitness of Cs is given by $\Pi_C(k) = \Pi_D(k) - c$.

The time evolution of the fraction of cooperators x in the population is given by the replicator equation,

$$\dot{x} = x(1 - x)(f_C - f_D).$$

It is straightforward to show that, for the NPD ($M = 0$), the right-hand side of the replicator equation will be positive (and hence, the fraction of cooperators will steadily increase) whenever $F > N$, since $f_C - f_D \sim (F/N) - 1$ (Appendix A). On the other hand, whenever $F < N$, $f_C - f_D < 0$ for $x \in [0, 1]$, and cooperators have no evolutionary chance.

Let us now consider the NSH, where $1 < M \leq N$. Let us assume that the return from the public good increases linearly with the number k of Cs, inasmuch as $k \geq M$. In view of the previous definitions, whenever $k < M$ no public good is produced, and hence Ds have a payoff of zero whereas Cs have a payoff of $-c$. The evolutionary dynamics of Cs and Ds in the NSH game with a minimum threshold M can again be studied by analysing the sign of $f_C - f_D$. We may write (see the electronic supplementary material)

$$f_C - f_D \equiv Q(x) = -c\left[1 - \frac{F}{N}R(x)\right],$$

where the polynomial $R(x)$ and its properties have been defined in Appendix A, whereas the details are provided in the electronic supplementary material. In a nutshell, the properties of $Q(x)$ lead to very interesting dynamics of the replicator equation, with possibly two interior fixed points (x_L and x_R, with $x_L \leq x_R$), as illustrated in Figure 7.1, for $N = 20$, different values of $1 < M \leq 20$ and variable F. Note, in particular, the fact that $R'(x_L) > 0$ and $R'(x_R) < 0$ (electronic supplementary material) allows us to classify immediately x_L as an unstable fixed point whereas x_R, if it exists, corresponds to a stable fixed point, as also illustrated in Figure 7.1. Moreover, when $(F/N) = R(M/N)$, M/N is the unique interior and unstable fixed point.

Between these two limiting values of F, and given the nature of the interior fixed points x_L and x_R, one can easily conclude that below x_L all individuals will ultimately forgo the public good. Conversely, for all $x > x_L$, the population will evolve towards a mixed equilibrium defined by x_R, corresponding to a stable fixed point of the associated replicator equation (even if, initially, $x > x_R$). "Three in a boat" provides the simplest possible case of this scenario. Similar to the NPD, whenever $F/N < R(M/N)$, $f_C(x) < f_D(x)$ for all x, which means that all individuals will end up forgoing the public good.

(b) Evolutionary dynamics in finite populations

Let us focus on a well-mixed population of size Z in the absence of mutations. Sampling of individuals is no longer binomial, following a hypergeometric distribution (see Appendix B). The fraction of cooperators is no longer a continuous variable, varying in steps of $1/Z$.

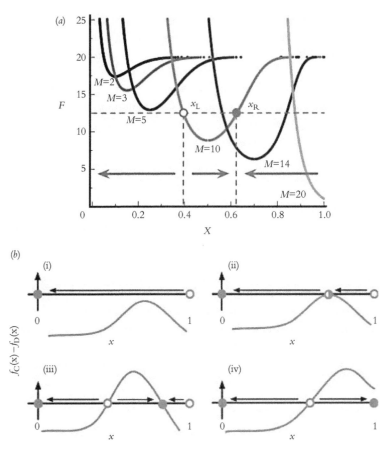

Figure 7.1 (a) Interior fixed points of the replicator equation for NSH games. The curves provide the location of the critical values of the fraction of cooperators ($x^* = \{x_L, x_R\}$) at which $f_C(x^*) = f_D(x^*)$. For each value of F (defining a horizontal line), the x^* values are given by the intersection of this line with each curve (one curve for given, fixed M, $N = 20$). Scenarios with none, one, and two interior fixed points are possible as detailed in (b(i–iv)): dynamics of NSH in infinite populations. Open circles represent unstable fixed points; filled circles represent stable fixed points and arrows indicate the direction of evolution by natural selection. For each case, the solid curves represent the typical shape of the function $f_C(x) - f_D(x)$. The quantity $\lambda^* = R(M/N)$ is defined in Appendix A and corresponds to the value of F at which the minimum of each curve in (a) for fixed M is reached. (i) $F/N < \lambda^*$, (ii) $F/N = \lambda^*$, (iii) $\lambda^* < F/N < 1$ and (iv) $F/N > 1$.

We adopt a stochastic birth–death process (Karlin and Taylor 1975) combined with the pairwise comparison rule (Traulsen et al. 2006, 2007a, 2007b) in order to describe the evolutionary dynamics of Cs (and Ds) in a finite population. Under pairwise comparison, two individuals from the population, A and B, are randomly selected for update (only the selection of mixed pairs can change the composition of the population). The strategy of A will replace that of B with a probability given by the Fermi function (from statistical physics),

$$p \equiv \frac{1}{1 + \exp(-\beta(f_A - f_B))}.$$

The reverse will happen with probability $1 - p$. The quantity β, which in physics corresponds to an inverse temperature, controls the intensity of selection: for $\beta \ll 1$, selection is weak, and one recovers the replicator equation in the limit $Z \rightarrow \infty$ (Traulsen et al. 2006, 2007a, 2007b). The pairwise comparison rule is similar to the so-called logit rule (Sandholm 2010), according to which an individual A is selected with a probability proportional to $e^{f_A/\eta}$; here the noise parameter η plays the role of the temperature above; in fact, both processes share the same fixation probabilities, despite leading to different evolutionary dynamics equations.

For arbitrary β, the quantity corresponding to the right-hand side of the replicator equation, specifying the "gradient of selection," is given in finite populations by (Traulsen et al. 2006, 2007a, 2007b)

$$g(k) \equiv T^+(k) - T^-(k) = \frac{kZ-k}{Z Z}\tanh\left\{\frac{\beta}{2}[f_C(k) - f_D(k)]\right\}. \qquad 2.1$$

The right-hand side of $g(k)$ is similar to the replicator equation, only the pairwise comparison leads to the appearance of the hyperbolic tangent of the fitness difference, instead of the fitness difference. This has implications in the characteristic evolutionary times, which now depend on β (Traulsen et al. 2006, 2007a, 2007b), but not in what concerns the roots of $g(k)$. Importantly, the evolutionary dynamics in finite populations will only stop whenever the population reaches a monomorphic state ($k/Z = 0$ or $k/Z = 1$). Hence, the sign of $g(k)$, which indicates the direction of selection, is important in that it may strongly influence the evolutionary time required to reach any of the absorbing states.

Whenever $M = 0$ (NPD) we may write (see appendix B)

$$f_C(k) - f_D(k) = c\left[\frac{F}{N}\left(1 - \frac{N-1}{Z-1}\right) - 1\right], \qquad 2.2$$

which is independent of k being, however, *population and group size dependent*. This means *frequency independent* selection. In particular, whenever the size of the group equals the population size, $N = Z$, we have $f_C(k) - f_D(k) = -c$ and cooperators have no chance irrespective of the value of the enhancement factor. This contrasts with the result in infinite, well-mixed populations ($Z \to \infty$), where to play C would be the best option whenever $F > N$. For finite populations, the possibility that group size equals population size leads to the demise of cooperation.

Given the independence of $f_C - f_D$ on k in finite populations, for a given population size there is a critical value of F for which selection is neutral, and above which cooperators will win the evolutionary race. From the two equations above, this critical value reads

$$F = N\left(1 - \frac{N-1}{Z-1}\right)^{-1}.$$

In Figure 7.2, we show the Z-dependence of $g(k)$ for fixed group size $N = 10$ and fixed $F = 12$ leading to a critical population size $Z = 55$.

Let us now discuss the NSH with $1 < M < N < Z$. Whenever $N = Z$, the result is easily inferred from the NPD above—all individuals in the population will ultimately forgo the public good. This will happen, in finite populations, irrespective of the existence (or not) of a threshold M. Whenever $N < Z$ the threshold brings about a strong disruption of the finite population dynamics, which we illustrate numerically, given the unappealing look of the analytical equations (see Appendix B).

Let us start with the case in which $F > N$, that is, the regime for which we obtain a pure coordination game with a single (unstable) fixed point in the replicator dynamics equation (cf. Figure 7.1). The possible scenarios are depicted in Figure 7.3a.

Clearly, for small population sizes, cooperators are always disadvantageous. With increasing Z, however, one approaches the replicator dynamics scenario (coordination game), despite the fact that, for example for $Z = 20$, convergence towards the absorbing state at 100 percent Cs is hindered because Cs become disadvantageous for large k. Indeed, for this population size, Cs are advantageous only in a small

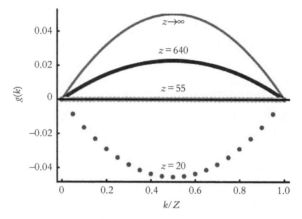

Figure 7.2 Behavior of an $g(k)$NPD game in which $F > N$. We plot $g(k)$ as a function of the (discrete) frequency of cooperators k/Z, for different values of the population size Z as indicated. Given that $F = 12$ and $N = 10$, for $Z = 55$, $g(k) = 0$ for all k, as depicted. Hence, selection is neutral and evolution proceeds via random drift, which means that the fixation probability of k Cs (or Ds) is simply k/Z. For values of Z below $Z = 55$, Cs are disadvantageous, whereas for values above $Z = 55$ Cs become advantageous, irrespective of the initial fraction of Cs initially present in the population, which corresponds to the evolutionary dynamics scenario in an infinite, well-mixed population.

neighborhood of $k/Z = 0.5$, being disadvantageous both for smaller and larger values of k/Z. In other words, and despite the fact that evolution will stop only at $k = 0$ or $k = Z$, the time it takes to reach an absorbing state will depend sensitively on the population size, given the occurrence (or not) of interior roots of $g(k)$.

Whenever $F < N$, yet above the critical limit below which Cs become disadvantageous for all x in Figure 7.1, we observe that for small population sizes Cs are always disadvantageous, and the two interior fixed points of the replicator dynamics equation only manifest themselves above a critical population size Z_{CRIT}, as illustrated in Figure 7.3b.

3. Discussion

In this chapter, we extend the range of PGG to systems where a minimum of coordinated collective action is required to achieve a public good. By doing so, we generalized the two-person SH game to N-person

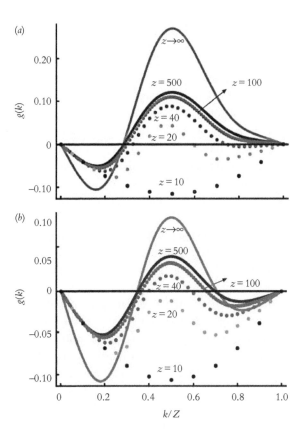

Figure 7.3 Behavior of $g(k)$ for an NSH game in a population of variable size Z and fixed group size $N = 10$, and $M = 5$. (a) Since $F = 12 > N$, the game becomes a pure coordination game in infinite populations. In finite populations, however, it strongly depends on Z: for $Z = N$, Cs are always disadvantageous and evolutionary dynamics leads mostly to 100 per cent Ds. For $Z = 20$ (and using a terminology which is only correct for $Z \rightarrow \infty$), we obtain a profile for $g(k)$ evidencing the emergence of a coordination point and a coexistence point. For increasingly large Z (e.g. $Z = 40$), the coexistence "point" disappears and we recover the behavior of the replicator dynamics: selection favors Cs above a given fraction k/Z and Ds below that fraction which, in turn, depends on the population size. (b) Since $F = 8 < N$, the game now exhibits two interior fixed points in infinite populations (dark grey curve). Similar to (a), for small Z Cs are disadvantageous for all k. Unlike (a), however, now the "interior fixed points" emerge together for a critical population size, and remain for larger population sizes.

games. In infinite, well-mixed populations, the existence of a threshold opens the possibility for the appearance of two interior fixed points in the replicator equation. The one at the lower frequency of cooperators is always an unstable fixed point, which determines a threshold for cooperative collective action. The other, at a higher frequency of cooperators, is a stable fixed point, and hence determines the final frequency of cooperators in the population, assuming that the coordination threshold is overcome. Besides this most interesting regime, there are also the possible outcomes of no cooperation or of a pure coordination game with a threshold that depends sensitively on the minimum number of cooperators M in a group of N individuals required to produce any public good.

Once the simplifying assumption of an infinite population size is abandoned, the evolutionary dynamics of the NSH game is profoundly affected, mostly when the population size is comparable with the group size (see Table 7.1 for a summary). In this regime, one observes an overlap of the different scenarios obtained in infinite populations. Hence, for $Z = N$, cooperators are always disadvantageous, irrespective of the existence or not of a threshold. For $Z > N$, the direction of selection in a finite population is strongly size dependent. For fixed $F > N$, there is a critical value of Z_1, above which the interior roots of $g(k)$ emerge, which constitute the finite-population analogs of x_L and x_R in infinite populations (cf. Figure 7.1). Above a second critical value Z_2, x_R disappears, and one ends up with a coordination game. For $M < F < N$ and a small population size, that is, $F < N$ but yet above the critical value $\lambda^* = R(M/N)$ defined in Appendix A and the electronic supplementary material, cooperators are always disadvantageous; however, above a critical population size (Z_C) the interior roots of $g(k)$ emerge simultaneously and the evolutionary dynamics approach that observed in infinite populations. Finally, for $F < M$ cooperators have no chance irrespective of the population size. Such strong size dependence, with an impact that is stronger for smaller population sizes, can be directly traced back to the fact that, for smaller populations, the hypergeometric sampling of individuals into groups significantly deviates from binomial sampling. This, in turn, reflects the intuition that, in small populations, choices are reduced, and this must influence the overall evolutionary dynamics.

In this work we have always assumed that the benefit returned by the PGG scales linearly with the amount contributed. This need not be the

Table 7.1 Interior roots of $g(k)$ for the NSH. One distinguishes two groups of interior roots of $g(k)$ which depend on how $F(>M)$ compares with N. When $F > N$, one approaches the infinite population size limit indirectly, in the sense that there is a first population threshold Z_1 above which two interior roots emerge, one of them disappearing above a second threshold Z_2. This scenario contrasts with that associated with $M < F < N$, for which there is a threshold Z_C at which two interior roots emerge, smoothly approaching the infinite limit with increasing population size Z (we used $\tilde{x}_L$ and $\tilde{x}_R$ to distinguish the roots for finite populations from those defined for infinite population).

Z	$M < F < N$	Z	$N < F$
$N \leq Z < Z_C$	—	$N \leq Z < Z_1$	—
$N < Z_C < Z$	$\tilde{x}_L, \tilde{x}_R$	$N < Z_1 < Z < Z_2$	$\tilde{x}_L, \tilde{x}_R$
$Z \to \infty$	x_L, x_R	$N < Z_1 < Z_2 < Z$	$\tilde{x}_L$
		$Z \to \infty$	x_L

case, and it is possible to find examples in which a nonlinear return paradigm would be more appropriate. Hence, it will prove interesting to understand in which way deviations from a linear return will affect evolutionary game dynamics, mostly in finite populations. Work along these lines is in progress.

Appendix A. Replicator Dynamics in Infinite Populations

We assume an infinite, well-mixed population, where x denotes the fraction of Cs, and $(1 - x)$ the fraction of Ds. Groups of N individuals are sampled randomly from the population and engaged in an NSH game. As referred to in Section 1, the NSH requires a minimum threshold of $M > 1$ ($M \leq N$) individuals for a public good to be produced whereas the NPD is obtained whenever $M = 0$. As a result, the average fitness of Ds in this population (as usual, we identify here fitness with payoff) is given by

$$f_D(x) = \sum_{k=0}^{N-1} \binom{N-1}{k} x^k (1-x)^{N-1-k} \Pi_D(k), \qquad (A1)$$

whereas the average fitness of Cs is given by

$$f_C(x) = \sum_{k=0}^{N-1} \binom{N-1}{k} x^k (1-x)^{N-1-k} \Pi_C(k+1), \qquad (A2)$$

where $\Pi_C(k)$ ($\Pi_D(k)$) are the fitness of a C (D) in a group of N individuals, k of which are Cs. Random sampling of individuals leads to groups whose composition follows a binomial distribution. In an N-individual group with Cs the fitness of Ds is given by

$$\Pi_D(k) = \frac{kFc}{N} \theta(k-M),$$

and that of Cs by

$$\Pi_C(k) = \Pi_D(k) - c,$$

where the Heaviside step function $\theta(x)$ satisfies $\theta(x < 0) = 0$ and $\theta(x \geq 0) = 1$. Hence, each C pays a fixed cost when engaging in a PGG, and the value of the public good increases linearly with the number k of Cs, inasmuch as $k \geq M$. In view of the previous definitions, whenever $k < M$ no public good is produced, and hence Ds have a payoff of zero whereas Cs have a payoff of $-c$.

For the NPD ($M = 0$), we readily obtain from equations (A1) and (A2) that $f_C - f_D \sim (F/N) - 1$ and cooperation becomes the preferred option whenever $F > N$. Whenever $F < N$, $f_C - f_D < 0$ for $x \in [0, 1]$, and cooperators have no evolutionary chance.

Whenever $M > 1$ and $k < M$, the situation is similar to the NPD: Cs remain disadvantaged in mixed groups. Whenever $k \geq M$, some public good is produced and now $\Pi_D(k) = (kFc/N)$ whereas $\Pi_C(k) = \Pi_D(k) - c$.

The evolutionary dynamics of Cs and Ds in the NSH game with a minimum threshold M can be studied by analyzing again the sign of $f_C - f_D$. We may write

$$f_C(x) - f_D(x) \equiv Q(x) = -c\left[1 - \frac{F}{N}R(x)\right],$$

(see the electronic supplementary material) where

$$R(x) = x^{M-1}\left[\sum_{k=M}^{N-1} \binom{N-1}{k} x^{k-M+1}(1-x)^{N-1-k} + M\binom{N-1}{M-1}(1-x)^{N-M}\right].$$

Table 7.2 Nature and number of fixed points of replicator dynamics. Given the definition of $\lambda^* = R(M/N)$, we identify the fixed points of the replicator dynamics, as well as their nature, for the different regimes associated with the possible values of the ratio F/N. Besides the trivial endpoints $\{0,1\}$, we also identify possible interior fixed points $\{x_L, x_R\}$ satisfying $x_L \in (0, M/N)$ and $x_R \in (M/N, 1)$ (see main text for additional details).

	$F/N < \lambda^*$	$F/N = \lambda^*$	$\lambda^* < F/N \leq 1$	$1 < F/N$
Stable	0	0	$0, x_R$	$0, 1$
Unstable	1	$M/N, 1$	$x_L, 1$	x_L

The roots of $Q(x)$ in $(0,1)$ determine whether the replicator equation exhibits interior fixed points. In the electronic supplementary material, we prove several properties of the polynomial $R(x)$. In particular, let us define $\lambda^* = R(M/N)$. Then (i) for $(F/N) < \lambda^*$ for there are no roots for $x \in (0, 1)$; (ii) for $(F/N) = \lambda^*$, M/N is a double root in $(0, 1)$; (iii) for $(F/N) < 1$, there is only one simple root $x_L \in (0, M/N)$; and (iv) whenever $\lambda^* < (F/N) \leq 1$ there are two simple roots $\{x_L, x_R\}$, with $x_L \in (0, M/N)$ and $x_R \in (M/N, 1)$. The implications of $R(x)$ in the evolutionary dynamics of the population are illustrated in Figure 7.1b and summarized in Table 7.2.

The fact that $R'(x_L) > 0$ and $R'(x_R) < 0$ allow us to classify immediately x_L as an unstable fixed point whereas x_R, if it exists, corresponds to a stable fixed point. Moreover, when $(F/N) = \lambda^*$, M/N is always an unstable fixed point.

Appendix B. Pairwise Comparison in Finite Populations

We consider now a finite well-mixed population of size Z, individual fitness resulting from engaging in an NSH. The average fitness of Cs and Ds now becomes a function of the (discrete) fraction k/Z of Cs in the population, and can be written as (hypergeometric sampling) (Hauert et al. 2007)

$$f_C(k) = \binom{Z-1}{N-1}^{-1} \sum_{j=0}^{N-1} \binom{k-1}{j}\binom{Z-k}{N-j-1} \Pi_C(j+1),$$

and

$$f_D(k) = \binom{Z-1}{N-1}^{-1} \sum_{j=0}^{N-1} \binom{k}{j}\binom{Z-k-1}{N-j-1} \Pi_D(j),$$

respectively, where we impose that the binomial coefficients satisfy $\binom{k}{j} = 0$ if $k < 0$.

We adopt a stochastic birth–death process (Karlin and Taylor 1975) combined with the pairwise comparison rule (Traulsen et al. 2006, 2007a, 2007b) introduced in Section 2b, in order to describe the evolutionary dynamics of Cs (and Ds) in a finite population. Given that we have k Cs in the population, the probability that, in a given time step, the number of Cs increases (decreases) by one is given by the transition probabilities

$$T^{\pm}(k) = \frac{k}{Z}\frac{Z-k}{Z}\frac{1}{1+\exp(\pm\beta[f_C(k)-f_D(k)])},$$

where β specifies the intensity of selection.

For finite populations, the quantity corresponding to the right-hand side of the replicator equation, specifying the "gradient of selection", is given by (Traulsen et al. 2006, 2007a, 2007b) $g(k)$ defined in equation (2.1) in Sectio 2b, and its interior roots are the roots of $f_C(k) - f_D(k)$. Since $\Pi_D(k) = (kFc/N)\theta(k - M)$ and $\Pi_C(k) = \Pi_D(k) - c$, we may explicitly write equation (2.2) of Section 2b for $f_C(k) - f_D(k)$ (see also the electronic supplementary material), whenever $M = 0$, *which is independent of k* being, however, *population and group size dependent*.

Whenever $M > 1$ and $Z = N$, the result is easily inferred from the NPD case. For $1 < M < N < Z$, the threshold brings about a strong disruption of the finite population dynamics, the analytical treatment of which is cumbersome. Numerically, however, the situation is easy to understand in light of the previous discussion. Consequently, Figure 7.3 was computed numerically using a direct implementation of the equations in *Mathematica*.

Acknowledgment

This work was supported by FCT Portugal (J.M.P.), FNRS Belgium (F.C.S.), and FAPERJ Brazil (M.O.S.).

References

Axelrod, R. (1984) *The Evolution of Cooperation*. New York: Basic Books.

Axelrod, R. and W. D. Hamilton (1981) "The evolution of cooperation." *Science* 211: 1390–6.

Beding, B. (2008) "The stone-age whale hunters who kill with their bare hands." *Daily Mail*, 12 April. See <http://www.dailymail.co.uk/news/article-465987/the-stone-age-whale-hunters-kill-bare-hands.html>.

Benaim, M., S. J. Schreiber, and P. Tarres (2004) "Generalized urn models of evolutionary processes." *Annals of Applied Probability* 14: 1455–78.

Boesch, C. (2002) "Cooperative hunting roles among Tai chimpanzees." *Human Nature* 13: 27–46.

Boyd, R. and P. J. Richerson (1985) *Culture and the Evolutionary Process*. Chicago, IL: University of Chicago Press.

Brandt, H., C. Hauert and K. Sigmund (2006) "Punishing and abstaining for public goods." *Proceedings of the National Academy of Science USA* 103: 495–7.

Bryant, J. (1994) "Coordination theory, the stag hunt and macroeconomics." In *Problems of Coordination in Economic Activity*, ed. J. W. Friedman, 207–25. Dordrecht, The Netherlands: Kluwer.

Creel, S. and N. M. Creel (1995) "Communal hunting and pack size in African wild dogs, *Lycaon pictus*." *Animal Behaviour* 50: 1325–39.

Hamilton, W. D. (1970) "Selfish and spiteful behaviour in an evolutionary model." *Nature* 228: 1218–20.

Hammerstein, P. (2003) *Genetic and Cultural Evolution of Cooperation*. Cambridge, MA: MIT Press.

Hauert, C., S. De Monte, J. Hofbauer, and K. Sigmund (2002) "Volunteering as Red Queen mechanism for cooperation in public goods games." *Science* 296: 1129–32.

Hauert, C., F. Michor, M. A. Nowak, and M. Doebeli (2006) "Synergy and discounting of cooperation in social dilemmas." *Journal of Theoretical Biology* 239: 195–202.

Hauert, C., A. Traulsen, H. Brandt, M. A. Nowak, and K. Sigmund (2007) "Via freedom to coercion: the emergence of costly punishment." *Science* 316: 1858.

Hofbauer, J. and K. Sigmund (1998) *Evolutionary Games and Population Dynamics*. Cambridge: Cambridge University Press.

Jervis, R. (1978) "Cooperation under the security dilemma." *World Politics* 30: 167–214.

Kandori, M., G. J. Mailath, and R. Rob (1993) "Learning, mutation, and long-run equilibria in games." *Econometrica* 61: 29–56.

Karlin, S. and H. M. A. Taylor (1975) *A First Course in Stochastic Processes.* London: Academic.

Kollock, P. (1998) "Social dilemmas: the anatomy of cooperation." *Annual Review of Sociology* 24: 183–214.

Macy, M. W. and A. Flache (2002) "Learning dynamics in social dilemmas." *Proceedings of the National Academy of Science USA* 99 (Suppl. 3): 7229–36.

Maynard-Smith, J. (1982) *Evolution and the Theory of Games.* Cambridge: Cambridge University Press.

Maynard-Smith, J. and E. Szathmáry (1995) *The Major Transitions in Evolution.* Oxford: Freeman.

Milinski, M., D. Semmann, H.J. Krambeck, and J. Marotzke (2006) "Stabilizing the Earth's climate is not a losing game: supporting evidence from public goods experiments." *Proceedings of the National Academy of Science USA* 103: 3994–8.

Milinski, M., R. D. Sommerfeld, H. J., Krambeck, F. A. Reed, and J. Marotzke (2008) "The collective-risk social dilemma and the prevention of simulated dangerous climate change." *Proceedings of the National Academy of Science USA* 105: 2291–4.

Nowak, M. A. (2006) "Five rules for the evolution of cooperation." *Science* 314: 1560–3.

Nowak, M. A. and K. Sigmund (2004) "Evolutionary dynamics of biological games." *Science* 303: 793–799.

Nowak, M. A., A. Sasaki, C. Taylor, and D. Fudenberg (2004) "Emergence of cooperation and evolutionary stability in finite populations." *Nature* 428: 646–50.8

Ohtsuki, H., C. Hauert, E. Lieberman, and M. A. Nowak (2006) "A simple rule for the evolution of cooperation on graphs and social networks." *Nature* 441: 502–5.

Packer, C. and T. M. Caro (1997) "Foraging costs in social carnivores." *Animal Behaviour* 54: 1317–18.

Rockenbach, B. and M. Milinski (2006) "The efficient interaction of indirect reciprocity and costly punishment." *Nature* 444: 718–23.

Sandholm, W. H. (2010) *Population Games and Evolutionary Dynamics.* Cambridge, MA: MIT Press.

Santos, F. C. and J. M. Pacheco (2005) "Scale-free networks provide a unifying framework for the emergence of cooperation." *Physical Review Letters* 95: 98–104.

Santos, F. C., J. M. Pacheco and T. Lenaerts (2006) "Evolutionary dynamics of social dilemmas in structured heterogeneous populations." *Proceedings of the National Academy of Science USA* 103: 3490–4.

Santos, F. C., M. D. Santos and J. M. Pacheco (2008) "Social diversity promotes the emergence of cooperation in public goods games." *Nature* 454: 213–16.

Schreiber, S. J. (2001) "Urn models, replicator processes, and random genetic drift." *SIAM Journal of Applied Mathemtics* 61: 2148–67.

Skyrms, B. (2001) The stag hunt. *Proceedngs and Addresses of the American Philosophical Association* 75: 31–41.

Skyrms, B. (2004) *The Stag Hunt and the Evolution of Social Structure.* Cambridge: Cambridge University Press.

Stander, P. E. (1992) "Cooperative hunting in lions: the role of the individual." *Behavioral Ecology and Sociobiology* 29: 445–54.

Sugden, R. (1986) *The Economics of Rights, Co-operation and Welfare.* Oxford: Basil Blackwell.

Taylor, M. and H. Ward (1982) "Chickens, whales, and lumpy goods: alternative models of public-goods provision." *Political Studies* 30: 350–70.

Traulsen, A., M. A. Nowak, and J. M. Pacheco (2006) "Stochastic dynamics of invasion and fixation." *Physical Review E: Statistical, Nonlinear, and Soft Matter Physics* 74: 011909.

Traulsen, A., M. A. Nowak, and J. M. Pacheco (2007a) "Stochastic payoff evaluation increases the temperature of selection." *Journal of Theoretical Biology* 244: 349–56.

Traulsen, A., J. M. Pacheco and M. A. Nowak (2007b) "Pairwise comparison and selection temperature in evolutionary game dynamics." *Journal of Theoretical Biology* 246: 522–9.

Ward, H. (1990) "Three men in a boat, two must row: an analysis of a three-person chicken pregame." *Journal of Conflict Resolution* 34: 371–400.

Young, H. P. (1993) "The evolution of conventions." *Econometrica* 61: 57–84.

8

Learning to Take Turns: The Basics

with Peter Vanderschraaf

1. Introduction

How do players learn to follow an equilibrium of a game?[1] Long before game theory was founded by von Neumann and Morgenstern (1944) and Nash (1950, 1951), David Hume suggested an answer: members of society gradually learn to follow a system of conventions regulating stable possession, or property rights, via "a slow progression, and by our repeated experience of the inconveniences of transgressing it" (1740: 490). In the early days of game theory, Brown (1951) developed a formal model of this sort of trial and error learning—*fictitious play*.[2] In traditional fictitious play, at successive time periods the players form predictions of the others' strategies based on historical frequency, and then play their best responses to their predictions.

Traditional fictitious play suffers from a serious limitation: players who learn by fictitious play cannot learn any nontrivial patterns of play. In particular, players who form their beliefs according to a traditional fictitious play process can never learn to *take turns* between outcomes that each favors. And yet when experimental game theorists have

---

[1] This paper introduces the leading ideas and main findings of Vanderschraaf and Skyrms (2003). For a careful treatment with precise definitions, proofs, comparisons of different games, details of simulations, and discussion of related literature please see that paper.

[2] For a different approach to learning patterns of play see Sosino (1997). For related variations on fictitious play see Fudenberg and Levine (1998).

subjects play repeated games that have "taking turns" equilibria, these subjects quickly learn to take turns. One might argue that the adults in game theory experiments take turns because they are already socialized. Taking turns is one of the techniques many people incorporate into their social tool kits.

But where does this technique come from in the first place? Here we use a simple learning dynamics, *Markov fictitious play*, to investigate the possibility of players spontaneously learning to take turns. As its name suggests, Markov fictitious play extends the traditional fictitious play process. Markov fictitious play is also perhaps the simplest form of adaptive dynamics based on pattern recognition. We find that players, starting from randomly chosen initial positions, who update their beliefs according to Markov fictitious play can learn to take turns quite often, although this result is by no means always bound to occur. Remarkably, such *Markov deliberators* frequently converge to "taking turns" equilibria not only when a pair play a game repeatedly with each other, but even when members of a large population with heterogeneous initial beliefs are matched at random at different times to play the game. We conclude that the simple Markov fictitious play processes introduced here show considerable promise as a learning model of the phenomenon of taking turns observed in the laboratory and in everyday life.

2. The Computer Game

Jan (Player 1) and Jill (Player 2) have a new Computer Game. Only one can play at a time. Playing is more fun than watching, but watching is more fun than doing nothing. If both try to play at the same time, they fight and no one gets to play. For a definite numerical realization, suppose that if one watches while the other plays the first gets a payoff of 3 and the second a payoff of 2. If the both attempt to play at the same time they both get a payoff of zero. If they both attempt to watch, they each get a payoff of 1.

If one plays then the other's unique best response is to watch, and conversely. So <Jan plays, Jill watches> and <Jill plays, Jan watches> are strict Nash equilibria of the game. This is an *impure coordination game* (Lewis 1969), because Jill's and Jan's preferences over these two strict equilibria conflict.

Traditional fictitious play leads to one of the following outcomes: Either players will settle into an inequitable sequence of plays where one always plays while the other always watches, so that the former never gets to enjoy her favorite outcome of the Computer Game. Or their actual sequence of plays oscillates between both trying to play and both trying to watch, so they miscoordinate on every play!

If rational agents start following a pattern of miscoordination, would they not notice this and try to break out of this pattern? Indeed, if a pair of agents like Jan and Jill are paired together to play the Computer Game repeatedly, one might expect them to settle into a pattern where they coordinate, not one where they miscoordinate. Specifically, one might expect them to take turns between the strict equilibria. This pattern of plays is not an equilibrium of the one-shot game, but is an equilibrium of the repeated game. By following a "taking turns" equilibrium, the players never miscoordinate on either of the suboptimal outcomes. In this situation, taking turns seems the best mechanism for sharing the benefits of coordination equally. And one might argue that taking turns serves as a prototype for more general forms of reciprocity.

Experimental game theory confirms that people, placed in situations where taking turns produces a fair and optimal equilibrium, do in fact tend to take turns. Faced with repeated impure coordination problems in the laboratory, subjects quickly settle into "taking turns" equilibria (Rapoport et al. 1976: chs 9, 10, 11; Prisbey 1992). Traditional fictitious play cannot account for this commonplace phenomenon. But the fictitious play model can be extended in a simple way that enables players to learn to take turns.

3. Markov Fictitious Play

Traditional fictitious play tacitly assumes that sample frequencies are the only information relevant to a player's prediction. This is why the traditional process cannot detect patterns. Moreover, traditional fictitious play uses an inductive logic designed for independent and identically distributed processes. We believe that learning models for game theory should be designed to accommodate sequences of play that are not independent and identically distributed, and in particular should be able to detect patterns of play.

The simplest patterns are ones in which probabilities at a time depend only on the state at the preceding time—a Markov chain. A minimal modification of fictitious play might then use an inductive logic designed for Markov chains (Kuipers 1988; Skyrms 1991). In our repeated game setting, the state of the system consists of the actions of both players. In *Markov Fictitious Play*, each player uses the counts of transitions to predict what the other will do this time, and then chooses the act which best responds to that prediction. Each player reasons as follows: In the last play I did this and he did that, and following such situations he chose acts with such and such frequencies. I use those frequencies as my predictions of what he will do next, and choose my immediate (myopic) best response.

4. Learning to Take Turns

Games played repeatedly by fixed players

Suppose Jan and Jill repeatedly play the Computer Game using *Markov Fictitious Play*. If they visit a strict Nash equilibrium twice in a row, then their play is "absorbed" into this strict equilibrium for all future rounds of play. But if they visit one strict Nash equilibrium, then visit another, and then return to the first, their play is "absorbed" into an alternating *taking turns* equilibrium of consecutive visits between the two strict equilibria. his result generalizes (Vanderschraaf and Skyrms 2003).

Alternating equilibria exist, but we would like to know whether they can arise spontaneously, and if so whether this should be an extremely unlikely occurrence or one that might occur naturally with some regularity. Suppose we just pick some initial beliefs for Jan and Jill at random, and start them out in Markov fictitious play. What should we expect to see?

We explored the properties of Markov fictitious play in the Computer Game by running a series of 10,000 computer simulations. In each simulation, a pair of initial belief matrices over the transitions of the states of the Computer Game were selected at random and then updated according to the Markov fictitious play rule. Beliefs nearly always converged to either a pure Nash equilibrium or an alternating equilibrium. In our simulations, players settled into an alternating equilibrium over a third of the time. (For details, see Vanderschraaf and Skyrms

2003.) We do not attach much significance to the exact numbers in these simulation results. But we can draw an important qualitative conclusion: From a randomly chosen starting point under Markov fictitious play, Jan and Jill can spontaneously learn to take turns. It is far from guaranteed that they will succeed, but it does not require a miracle for them to do so.

Games played repeatedly within populations

Suppose that Jan and Jill do not have the computer to themselves, but that other children in the room take their places, and are themselves replaced by others. All the children have seen what has happened, but each child has her own initial weights for transitions. In this social setting we have a far more heterogeneous assortment of learning styles, here represented by different inductive rules. Is it possible to learn to take turns in this more challenging setting?

We can apply Markov fictitious play in this social setting in a straightforward way. At each round fresh individuals assume the roles of each player and update according to the observed past history of plays. In this variant of Markov fictitious play, the dynamical system of updated beliefs is constantly bombarded by the random fluctuations in individual players' priors and weighting constants. It is always theoretically possible for newcomers with their own idiosyncratic weighting constants and priors to disrupt the system and throw the population off the incumbent equilibrium.

Do Markov deliberators in such an inherently "noisy" system have any prospect of reaching an equilibrium? As a matter of fact, for the 2×2 case this Markov dynamics exhibits remarkable convergence properties. Again we ran computer simulations, this time with each representative's initial transition probabilities and weighting factor picked at random. We were somewhat surprised to find that the results in this social learning model were quite similar to those of the previous case where Jan and Jill were the only players. In these simulations, taking turns emerges about a quarter of the time, and play *always* converges to either one of the strict Nash equilibria or to an alternating equilibrium. In these cases the expectation that the players would take turns arose spontaneously and then was passed along to new players in the social context.

5. Conclusion

Markov fictitious play can model the phenomenon of taking turns observed in the laboratory and in everyday life. Markov fictitious play can converge quite rapidly to an alternating equilibrium, which reflects the rapid convergence to taking turns observed in experiments (Rapoport et al. 1976). We do not claim that the Markov fictitious play processes analyzed here are accurate models of human learning. Markov fictitious play surely oversimplifies the learning processes that occur in human communities. Yet the very simplicity of this model is illuminating. Even a community of players who have no prior history of interactions, but who update their beliefs according to a simple Markov fictitious play process, can rapidly learn to take turns. We conjecture that players like Jan and Jill, who are neither so naive nor so ignorant as the deliberators of our Markov learning model, have even better prospects for learning to take turns.

References

Brown, G. W. (1951) "Iterative Solutions of Games by Fictitious Play." In *Activity Analysis of Production and Allocation*, ed. T. C. Koopmans, 374–6. New York: Wiley.

Fudenberg, D. and D. Levine (1998) *The Theory of Learning in Games*. Cambridge, MA: MIT Press.

Hume, D. (1740, 1888) *A Treatise of Human Nature*, ed. L. A. Selby-Bigge (1976), Oxford: Clarendon Press.

Kuipers, T. A. F. (1988) "Inductive Logic by Similarity and Proximity." In *Analogical Reasoning*, ed. D. A. Helman. Dordrecht: Kluwer Academic Publishers.

Lewis, D. (1969) *Convention: A Philosophical Study*. Cambridge, MA: Harvard University Press.

Nash, J. (1950) "Equilibrium Points in N-Person Games." *Proceedings of the National Academy of Sciences of the USA* 36: 48–9.

Nash, J. (1951) "Non-cooperative Games." *The Annals of Mathematics* 54: 286–95.

Prisbey, J. (1992) "An Experimental Analysis of Two-Person Reciprocity Games." Social Science Working Paper 787, California Institute of Technology.

Rapoport, A., M. Guyer, and D. Gordon (1976) *The 2 × 2 Game*. Ann Arbor: University of Michigan Press.

Skyrms, B. (1991) "Carnapian Inductive Logic for Markov Chains." *Erkenntnis* 35, 439–60.

Sonsino, D. (1997) "Learning to Learn, Pattern Recognition, and Nash Equilibrium." *Games and Economic Behavior* 18: 286–331.

Vanderschraaf, P. (2001) *Learning and Coordination.* New York: Routledge.

Vanderschraaf, P. and B. Skyrms (2003) "Learning to Take Turns." *Erkenntnis* 59: 311–48.

Von Neumann, J. and O. Morgenstern (1944) *Theory of Games and Economic Behavior.* Princeton, NJ: Princeton University Press.

9

Evolutionary Considerations in the Framing of Social Norms

with Kevin J. S. Zollman

In the implementation of social norms, context and framing effects can make enormous differences. This is commonplace in the rich social psychology literature that is brought to bear in Cristina Bicchieri's (2006) *The Grammar of Society*.[1] She sees the application of norms as being controlled by social scenarios or *scripts* that people act out. The same problem may elicit different behavior depending on which script has been initiated and which norm it has activated.

The importance of such framing of a decision problem has been so well documented in experimental psychology and economics that experimentalists no longer spend much time trying to elicit it. Rather, they take great care in constructing experiments with 'neutral' protocols, designed to avoid framing the problem one way or another.

But we really need a theory that explains both the genesis of norms and the possibility of framing effects. What might the framework for such a theory be? As David Hume saw long ago, social norms arise by a slow process of cultural evolution:

Nor is the rule concerning the stability of possession the less deriv'd from human conventions, that it arises gradually, and acquires force by a slow progression,

---

[1] C. Bicchieri (2006) *The Grammar of Society: The Nature and Dynamics of Social Norms*. Cambridge: Cambridge University Press.

and by our repeated experience of the inconveniences of transgressing it.[2] (Hume 2003 [1739–49])

Evolutionary game theory is the correct locus of our concerns.

Most evolutionary analyses concentrate on specific games. It is clear, however, that we do not evolve separate norms for individual games. Social norms evolve in the context of broad classes of social interactions—in a variety of games. Furthermore, they evolve in a way conductive to ambiguity in application. If there were a partition of social interactions, with a separate norm evolving for each class in the partition, application would be straightforward. But the classes of interactions driving the application of norms need not have any nice structure, and the norms that evolve may well not do so either.

A specific kind of interaction of the kind that game theorists put under the microscope (for instance, ultimatum bargaining) may be a member of various classes of social interactions. Each of these classes may carry its own norm, quite appropriate to the class, and the norms may conflict. We suggest that the framing of a decision problem should be interpreted as a signal about the relevant class of social interactions, and that evolutionary analysis should be redirected to systems of classes of social interactions. In this way, many 'anomalous' findings of experimental game theory may lose their air of mystery. We illustrate the approach here with two small examples, which illustrate two different types of framing effects.

Individual behavior across strategic contexts can often be very similar. For instance, it is remarkable how stable bargaining behavior is across contexts that are, from an economic perspective, strategically very different. Fair bargaining outcomes, where individuals both receive an equal share of a good to be split, have been observed in many strategically different games—from the Nash bargaining game, where individuals each choose an amount to demand for themselves, to the ultimatum game, where one individual suggests a split of a good, to the dictator game, where one individual unilaterally decides how to split the good.

The latter two have been viewed as especially problematic. In the ultimatum game, one individual suggests a split of the good and the other is given the opportunity to accept or reject the split. If the proposer presumes that his counterpart is rational and prefers something to

---

[2] D. Hume (2003 [1739–40]) *A Treatise of Human Nature*. Mineola, NY: Dover, p. 348.

nothing, then the proposer expects that the responder will accept any positive offer, since she is choosing between something and nothing. With this presumption, the proposer maximizes his own return by suggesting a split that gives the responder the smallest positive amount possible. Strategy sets that survive this reasoning are known as sequentially rational equilibria, but experiments have demonstrated that often individuals do not even approximate sequentially rational play.

Even more perplexing are results from dictator games, where the option to refuse has been removed. In these games, a single individual determines a split and the second must take the offer. Here, if the first individual prefers more money to less, he does best by keeping all the money for himself. However, experiments again show that individuals do not behave in this way.

Turning from a traditional equilibrium analysis to an evolutionary one does not entirely remove the mystery. While fair behavior in the ultimatum game can evolve in replicator dynamics (one model of cultural evolution), it is unstable to many types of mutation. Even considering a limited form of mutation, standard evolutionary models make the evolution of fair behavior relatively unlikely.[3] In the dictator game, it cannot evolve at all.

If one is to explain this phenomenon, it seems natural to suppose that individuals are relying on a single norm of bargaining when deciding what to do in many different bargaining-like circumstances, and that this norm does not differentiate between different strategic contexts. This suggestion has been echoed many times, perhaps most clearly by Gale, Binmore, and Samuelson:

In particular we suggest that initial play reflects decision rules that have evolved in real-life bargaining situations that are superficially similar to the Ultimatum Game. These bargaining games generally feature more symmetric allocations of bargaining power than the Ultimatum Game, yielding initial play in the Ultimatum Game experiments that need not be close to [sequentially rational play].[4] (Binmore et al. 1995: 59; see also Skyrms 1996)

---

[3] For a detailed study of evolutionary dynamics in the ultimatum game, see K. Binmore et al. (1995) and Harms (1997).

[4] Binmore et al., "Learning to be Imperfect." p. 59. See also B. Skyrms (1996) *Evolution of the Social Contract*. Cambridge: Cambridge University Press.

The evolutionary question then is "Is it possible for a general norm to evolve, such that the norm results in fair behavior both in the ultimatum and Nash bargaining games?" This question was investigated by Zollman (2008).[5] He considers a single norm, which dictates both the proposal in the symmetric Nash bargaining game (where both individuals simultaneously make demands) and in the asymmetric ultimatum game. He finds that when individuals cannot distinguish between these two circumstances, fair behavior is very likely to evolve. More surprisingly, he finds that there is a synergistic effect, and that fair behavior is more likely to evolve in the generic context than it is in *either* game taken alone.

Zollman assumes that the fairness norm in question simply cannot distinguish between the two games, thus forcing it to evolve consistent behavior in both contexts. He defends this constraint by saying

Individuals may simply fail to consider the full strategic situation with which they are confronted. Costa-Gomes, Crawford, and Broseta provide an extensive study demonstrating that a reasonable number of individuals simply do not consider game-like situations strategically.... Even for those players who would consider the actions of another, in many bargaining situations the information may be strictly unavailable to the players (or at least unavailable at a reasonable expense).[6]

The final remark suggests that it may be too expensive for individuals to make distinctions between different forms of the game, and thus they must adopt a strategy which is not game contingent. This possibility is analyzed in detail by Mengel (2008).[7] Mengel considered a situation where individuals confront many games, but incur some cost for distinguishing between them. She finds that depending on cost, many strategically very different situations can arise. New equilibria can be created and other equilibria destabilized. This suggests that the phenomenon found by Zollman may be a general feature of the evolution of norms.

These contributions model a situation where two strategically different interactions are coalesced into a single socially relevant context for an individual. Some players may frame ultimatum bargaining as just a

[5] K. J. S. Zollman (2008) "Explaining Fairness in Complex Environments." *Politics, Philosophy and Economics* 7: 81–97.

[6] Zollman (2008: 89). The study by Costa-Gomes et al. referred to in the quote is Costa-Gomes, Crawford, and Broseta (2001).

[7] F. Mengel (2008) "Learning Across Games." IVIE Working Paper AD 2007–05, <http://merlin.fae.ua.es/friederike/Dateien/LAGjan09.pdf>.

bargaining problem, some as a very special kind of bargaining, with different behavior as a result.

This is not the only time that framing effects matter, it can also be the case that individuals behave differently in one strategic context by cuing on some strategically irrelevant piece of information. For instance, it has been observed that in bargaining games apparently irrelevant bits of information can radically alter the agreed upon outcome. Although many such experiments have been performed, perhaps the most incredible results come from Mehta, Starmer, and Sugden (1992).[8] In this experiment, two individuals were randomly given four cards each from a reduced deck of cards containing only aces and deuces. Mehta et al. found that the focal point created by the cards resulted in asymmetric demands in the Nash bargaining game. They cued their subjects in such a way as to suggest that aces were valuable, and so individuals with more aces tended to demand more and, symmetrically, individuals with fewer demanded less. This shows people coordinating on different equilibria by cuing on strategically irrelevant information.

Let us move to a famous example discussed in the study of biological evolution in animal contests. Maynard Smith (1982) suggested that animal conflicts might be modeled using a game known as "Hawk–Dove."[9] In this game, each individual would like to secure a resource (such as food or a mate) and can do so by threatening the other. But if both threaten, they will fight, which is the worst outcome of the game.

Table 9.1 provides the payoff matrix for this game. In this game, there are two pure strategy Nash equilibria where one person plays *Dove* and the other *Hawk*. Such equilibria are attainable when there is an

Table 9.1 Hawk-Dove

	Hawk	Dove
Hawk	(0,0)	(4,1)
Dove	(1,4)	(2,2)

[8] J. Mehta, C. Starmer, and R. Sugden (1992) "An Experimental Investigation of Focal Points in Coordination and Bargaining: Some Preliminary Results." In *Decision Making Under Risk and Uncertainty: New Models and Findings*, ed. J. Geweke. Norwell: Kluwer.

[9] John Maynard Smith (1982) *Evolution and the Theory of Games*. Cambridge: Cambridge University Press.

0   p' 1

Figure 9.1 One-population replicator dynamics for the Hawk-Dove game

asymmetry available to the players (such as one being designated the 'row player'). In natural interactions where strategies correspond to types in the population, these equilibria are no longer attainable.

This can be illustrated with a relatively simple model of evolution known as "replicator dynamics." In the replicator dynamics model, individuals play with a random member of a population and reproduce according to how well they do relative to the rest of the population; those that do better than average take over larger shares of the population, while the shares of those who do worse shrink. When the situation is entirely symmetrical in the Hawk–Dove game, evolution pushes the population toward a mixed state where some proportion play Hawk and some play Dove (see Figure 9.1).

This polymorphic equilibrium is inefficient, despite being the unique endpoint for this model of evolution. Hawk-types often meet other Hawk-types and have the worst payoff in the game, zero. Dove-types also meet other dove-types, which results in a worse social outcome than would be secured if one had played Hawk (a sum of four instead of five). Maynard-Smith noted that in such games it might be of interest to the players to find something outside of the game to use as a method for breaking symmetry. That is, if the players could use some feature observable to both players to which they can correlate their strategy, evolution might select for strategies which use this cue.

This model has been used to explain territoriality in many species. Being a "territory owner" or an "intruder" is a mechanism by which individuals might correlate their strategies, and thus solve this coordination problem. So the strategy of playing *Hawk if owner and dove if intruder* is evolutionarily stable and also more efficient than the mixed population mentioned before. We can think of individuals sometimes being in the role of owner and sometimes in the role of intruder, with nature determining the roles with coin flips and matching owners with intruders.

Individuals can now evolve role-based strategies of the form <do this if owner, do that if intruder>. There are four such strategies (<H,H>, <D,D>, <H,D>, and <D,H>), so the dynamics live on a tetrahedron of

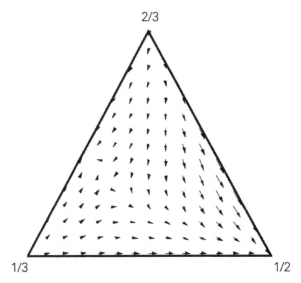

2/3

1/3   1/2

Figure 9.2 Evolutionary dynamics of the Nash bargaining game

population proportions. If <H,D> and <D,H> are extinct, players ignore their roles and we are back in the previous case with a unique polymorphic equilibrium. But when we consider the whole tetrahedron, this equilibrium becomes unstable. If there are a few of the other types (<H,D> and <D,H>) in the population and their proportions are not exactly equal, evolutionary dynamics will move away from this equilibrium. There are lots of new equilibria created by the introduction of roles, but many are also unstable. The bottom line is that now almost every population state is carried by evolution to either all <H,D> (Maynard-Smith's 'Bourgeois' strategy) or all <H,D> (Maynard-Smith's 'Paradoxical' strategy).

To describe the same thing in different terms, nature sends a signal to individuals, the signals are (anti-)correlated, and individuals have strategies that are conditional on the signal. In the population states all <H,D> or all <D,H>, we have realizations of a special case of what Aumann (1974) calls a *correlated equilibrium*.[10] Now let us bring the discussion back to context dependence. For situations where the role is

---

[10] R. Aumann (1974) "Subjectivity and Correlation in Randomized Strategies." *Journal of Mathematical Economics* 1: 67–96.

unclear (an ambiguous signal or no signal at all), we should expect a polymorphism of Hawks and Doves. For situations with a clear signal, we should expect a correlated equilibrium.

The situation of Hawk–Dove is very similar to the symmetry observed in bargaining games. In the Nash bargaining game, there are both symmetric and asymmetric equilibria. There are two symmetric equilibria, one being the fair equilibrium where each person demands half of the good and the other being the greedy equilibrium where both demand all of the good. There are also infinitely many asymmetric equilibria where one person demands x percent of the good and the other demands 100 – x percent of the good.

Restricting our attention to a game with just three strategies, which are demand one-third, demand half, and demand two-thirds, we find there are three equilibria: one where the first person demands one-third and the second demands two-thirds; one where the first person demands two-thirds and the second demands one-third; and one where both demand half. If we consider the one-population replicator dynamics again, we find that most populations go to the symmetric equilibrium, but a not insignificant number go to an inefficient mixed equilibrium where the population is made up of both one-third-types and two-thirds-types.[11] This mixed state is inefficient, as in the Hawk–Dove case, because sometimes two-thirds-types meet themselves and get nothing and sometimes one-third-types meet themselves and leave some of the good unused.

Suppose, however, that players receive a signal that is blue or green (B or G). Then there are available new strategies, such as <one-third if B, two-thirds if G>, and so forth. Suppose that nature assigns roles with a coin flip and matches blues with greens. Then, just as before, the polymorphism between the types always one-third and always two-thirds becomes destabilized by the addition of these new role-based strategies. We now have new stable equilibria where the whole population plays <one-third if B, two-thirds if G> and where the whole population plays <two-thirds if B, one-third if G>. We also have a stable equilibrium where the whole population demands half, regardless of signal.

One respect in which cultural evolution differs from biological evolution is with regard to the method of transmission. Although each

---

[11] Skyrms (1996: fig. 3). See Figure 9.2 in the present chapter.

individual in our bargaining game is endowed with a contingency plan for each circumstance (what to do if blue and what to do if green), this need not be transmitted in whole to the next "generation" of strategies. Individuals might imitate one person's contingency plan for blue and another person's contingency plan for green if they are differentially successful in the different contexts.

This possibility is modeled by a slight variant of the replicator dynamics discussed above. Instead of having a single population with full contingency plans evolving, we can treat the different contingency plans as evolving separately from one another. If blues always play against greens, this results in a model of evolution known as 'two-population replicator dynamics'. What-to-do-if-blue strategies compete against other what-to-do-if-blue strategies, and similarly for contingency plans for green.

In this model, we find that asymmetric equilibria become more likely. Considering a game with only three proposals (one-third, half, and two-thirds), almost half of the initial starting points evolve to situations where one type demands two-thirds and the other demands one-third. The remaining populations evolve to a state where both demand half.

We now have interesting possibilities for framing depending on whether signals are received or not. That is to say that we can evolve one norm in the absence of signals and other norms in the presence of signals. The asymmetries between social classes come to mind, where division may be egalitarian within classes, but inegalitarian between classes. Since strategically irrelevant properties can break the symmetry, we may see how the process of formation of social classes may originate.[12]

Signals can indicate the structure of the underlying game, as in our first example of Nash and ultimatum bargaining. No-cue behavior may be different from cued behavior. Signals may break symmetry and enable correlated equilibria in a single underlying game, as in our other examples. No-cue behavior again may be different from cued behavior. More generally, the point can be generalized to more complex systems of signals for more complex classes of games, which may combine these effects.

There is nothing very surprising in our general point, and we think that most experimentalists would agree. For instance, in a review of the literature on repeated Stag Hunt games, Van Huyck, Battalio, and

---

[12] For a similar observation in a spatial model, see Axtell et al. (1999).

Rankin (1997) find that on average, subjects walk into the laboratory with an inclination toward the payoff-dominant solution (stag hunting), but learn to follow risk dominance (hare hunting) as the experiment is repeated.[13] Obviously, the norms that subjects initially use are not formed by repetitions of the same interaction used in the experiment, but by some broader class of interactions.

In an intriguing follow-up experiment by Rankin, Van Huyck, and Battalio (2000), subjects played a sequence of different stag hunt games without labels, so that the only information they have is payoff informa-tion.[14] Three out of four groups coordinated on a norm of following the payoff-dominant equilibrium and the fourth group appeared on the way to converging on this norm. This unique experiment points the way to a more general type of experiment relevant to the framing of norms.

On a larger scale, some norms may be formed by classes of inter-actions so broad as to be thought of as constituting a culture. We take this as the moral of the multicultural studies in Henrich et al. (2000).[15] These shatter a lot of universal images—not only that of *homo econom-icus*, but also alternatives formed in experiments on college students in modern developed countries. The way individuals play standard games varies dramatically from one culture to another. It is evident that other kinds of customary social interactions in the culture play a role in shaping norms that are applied in public goods provision games or ultimatum bargaining. The challenge of studying the origin and framing of systems of norms has implications for experimental econom-ics, but it also calls for empirical research outside the laboratory.

References

Aumann, R. (1974) "Subjectivity and Correlation in Randomized Strategies." *Journal of Mathematical Economics* 1: 67–96.

[13] J. B. Van Huyck, R. C. Battalio, and F. W. Rankin (1997) "On the Origin of Convention: Evidence from Coordination Games." *Economic Journal* 107: 576–96.

[14] F. W. Rankin, J. B. Van Huyck, and R. C. Battalio (2000) "Strategic Similarity and Emergent Conventions: Evidence from Similar Stag Hunt Games." *Games and Economic Behavior* 32: 315–37.

[15] J. Henrich, R. Boyd, R. Bowles, C. Camerer, E. Fehr, and H. Gintis (2004) *Foundations of Human Sociality: Economic Experiments and Ethnographic Evidence from Fifteen Small-Scale Societies.* Oxford: Oxford University Press.

Axtell, R., J. Epstein, and H. P. Young (1999) "The Emergence of Classes in a Multi-Agent Bargaining Model." In *Social Dynamics*, ed. S. Durlauf and H. P. Young. Oxford: Oxford University Press.

Bicchieri, C. (2006) *The Grammar of Society: The Nature and Dynamics of Social Norms.* Cambridge: Cambridge University Press.

Binmore, K., J. Gale, and L. Samuelson (1995) "Learning to be Imperfect: The Ultimatum Game." *Games and Economic Behavior* 8: 56–90.

Costa-Gomes, M., V.P. Crawford, and B. Broseta (2001) "Cognition and Behavior in Normal-Form Games: An Experimental Study." *Econometica* 5: 1193–235.

Harms, W. (1997) "Evolution and Ultimatum Bargaining." *Theory and Decision* 42: 147–75.

Henrich, J., R. Boyd, R. Bowles, C. Camerer, E. Fehr, and H. Gintis (2004) *Foundations of Human Sociality: Economic Experiments and Ethnographic Evidence from Fifteen Small-Scale Societies.* Oxford: Oxford University Press.

Hume, D. (2003 [1739–40]) *A Treatise of Human Nature.* Mineola, NY: Dover.

Mehta, J., C. Starmer, and R. Sugden (1992) "An Experimental Investigation of Focal Points in Coordination and Bargaining: Some Preliminary Results." In *Decision Making Under Risk and Uncertainty: New Models and Findings*, ed. J. Geweke. Norwell: Kluwer.

Mengel, F. (2008) "Learning Across Games." IVIE Working Paper AD 2007-05, <http://merlin.fae.ua.es/friederike/Dateien/LAGjan09.pdf>.

Rankin, F. W., J. B. Van Huyck, and R. C. Battalio (2000) "Strategic Similarity and Emergent Conventions: Evidence from Similar Stag Hunt Games." *Games and Economic Behavior* 32: 315–37.

Skyrms, B. (1996) *Evolution of the Social Contract.* Cambridge: Cambridge University Press.

Smith, J. M. (1982) *Evolution and the Theory of Games.* Cambridge: Cambridge University Press.

Van Huyck, J. B., R. C. Battalio, and F. W. Rankin (1997) "On the Origin of Convention: Evidence from Coordination Games." *Economic Journal* 107: 576–96.

Zollman, K. J. S. (2008) "Explaining Fairness in Complex Environments." *Politics, Philosophy and Economics* 7: 81–97.

PART III

Dynamic Networks

Introduction

In 1993, at the Nobel Symposium on Game Theory, Al Roth gave a paper on low rationality learning in game theory using reinforcement learning. I had already come to the conclusion that common knowledge assumptions of high rationality game theory were unrealistically strong, and I liked this paper very much. I was interested in modeling game play on dynamically evolving social networks, and decided to investigate networks evolving by reinforcement learning. Except for one trivial case, the best I could do was to run extensive simulations but they produced very interesting results. I showed these to my friend Persi Diaconis. He had invented *reinforced random walks*, which use the same basic dynamics as Roth and Erev. The appropriate mathematical tool is stochastic approximation theory, of which I was then entirely innocent. Persi introduced me to his student Robin Pemantle, an expert in the field. All the chapters in this section and one in the next are products of our collaboration.

The first paper "Learning to Network" was written last, and provides an overview of the conclusions of the others. It should be easy to read. The remaining papers deliver the goods.

There is a lot of recent work on the co-evolution of network structure and strategy, both empirical—in humans and animals—and theoretical. A sampling is referenced below.

References

Cantor, M. and H. Whitehead (2013) "The Interplay between Social Networks and Culture: Theoretically and among Whales and Dolphins." *Philosophical Transactions of the Royal Society B* 368.

Cao, L., H. Otsuki, B. Wang, and K. Aihara (2011) "Evolution of Cooperation in Adaptively Weighted Networks." *Journal of Theoretical Biology* 272: 8–15.

Chiang, Y. S. (2008) "A Path Toward Fairness: Preferential Association and the Evolution of Fairness in the Ultimatum Game." *Rationality and Society* 20: 173–201.

Fosco, C. and F. Mengel (2011) "Cooperation through Imitation and Exclusion in Networks." *Journal of Economic Dynamics and Control* 35: 641–58.

Fowler, J., and N. Christakis (2010) "Cooperative Behavior Cascades in Human Social Networks." *PNAS* 107: 5334–8.

Gross, T. and B. Blasius (2008) "Adaptive Coevolutionary Networks: A Review." *Journal of the Royal Society Interface* 5: 259–71.

Gross, T. and H. Sayama (2009) *Adaptive Social Networks*. Berlin: Springer.

Rand, D., S. Arbesman, and N. Christakis (2011) "Dynamic Networks Promote Cooperation in Experiments with Humans." *PNAS* 108: 19193–8.

Santos, F. J., J. Pacheco, and T. Lenaerts (2006) "Cooperation Prevails when Individuals Adjust their Social Ties." *PLoS Computational Biology* 2 (10): e140.

Song, Z. and M. W. Feldman (2013) "The Coevolution of Long-term Pair Bonds and Cooperation." *Journal of Evolutionary Biology* 26: 963–70.

Spiekermann, K. (2009) "Sort out your Neighborhood: Public Goods Games on Dynamic Networks" *Synthese* 168: 273–94.

Wang, J., S. Suri, and D. J. Watts (2011) "Cooperation and Assortativity with Dynamic Partner Updating." *PNAS* 109: 14363–8.

10

Learning to Network

with Robin Pemantle

1. Introduction

In species capable of learning, including our own, individuals can modify their behavior by some adaptive process. Important classes of behavior—mating, predation, coalitions, trade, signaling, and division of labor—involve interactions between individuals. The agents involved learn two things: with *whom to interact* and *how to act*. That is to say that adaptive dynamics operates both on structure and strategy.

In an interaction, individuals actualize some behavior, the behavior of the individuals jointly determines the outcome of the interaction, and the consequences for the individuals motivate learning. At this high level of abstraction, we can model interactions as games. The relevant behaviors of individuals are called strategies of the game, and the strategies of the players jointly determine their payoffs. Payoffs drive the learning dynamics (Skyrms and Pemantle 2000).

If we fix the interaction structure in this abstract scheme, we get models of the evolution of strategies in games played on a fixed structure. An interaction structure need not be deterministic. In general, it can be thought of as a specification of the probabilities of interaction with other individuals. By far the most frequently studied interaction structure is one in which the group of individuals is large and individuals interact at random. That is to say that each individual has equal probability of interacting with every other individual in the population. Among a list of virtues of this model, mathematical tractability must come near the top. At another end of the spectrum we have models where individuals interact with their neighbors on a torus, or a circle, or (less frequently) some other graphical structure (Nowak and May 1992; Ellison 1993; Hegselmann 1996; Alexander 2000). Except in the simplest cases, these models sacrifice mathematical tractability to gain realism, and computer simulations have played an important role in their investigation. These two extreme models, however, can have quite different implications for the evolution of behavior. In large, random-encounter settings cooperators are quickly eliminated in interactions with a Prisoner's Dilemma structure. Comparable local interaction models allow cooperators to persist in the population.

If we fix the strategies of individuals and let the interaction structure evolve we get a model of interaction network formation. Evolution of structure is less well-studied than evolution of strategies, and is the main focus of this chapter. Most current research on the theory of network formation takes the point of view that networks are modeled as graphs or directed graphs, and network dynamics consists of making and breaking of links. (Bala and Goyal 2000; Jackson and Watts 2002). In taking an interaction structure to be a specification of probabilities of interaction rather than a graphical structure, we take a more general view than most of the literature (but see Kirman 1997 for a point of view close to that taken here). It is possible that learning dynamics may drive these probabilities to zero or one and that a deterministic graphical interaction structure may crystallize out, but this will be treated as a special case. We believe that this probabilistic approach can give a more faithful account of both human and non-human interactions. It also makes available a set of mathematical tools that do not fit the coarser picture of making or breaking deterministic links in a graphical structure.

Ultimate interest resides in the general case where structure and strategy co-evolve. These may be modified by the same or different kinds of learning. They may proceed at the same rate or different rates. The case where structure dynamics is slow and strategy dynamics is fast may approximate more familiar models where strategies evolve on a fixed interaction structure. The opposite case may be close to that of individuals with fixed strategies (or phenotypes) learning to network. In between, there is a very rich territory waiting to be explored. We will close this chapter with a discussion of the co-evolution of structure and strategy in a game that one of us has argued is the best simple prototype of the problem of instituting a social contract (Skyrms 2004). Whether co-evolution of structure and strategy supports or reverses the conventional wisdom about equilibrium selection in this game depends on the nature and relative rates of the two learning processes.

2. Learning

Learning can be divided into two broad categories: (1) belief learning in which the organism forms beliefs (or internal representations of the world) and uses these to make decisions, and (2) reinforcement learning, where the organism increases the probability of acts that have been rewarded and decreases the probability of those that have not been rewarded. Ultimately the distinction may not be so clear-cut, but it is useful for a categorization of learning theories. In the simplest belief learning model, Cournot dynamics, an individual assumes that others will do what they did last time and performs the act that has the highest payoff on that assumption. More sophisticated individuals might form their beliefs more carefully, by applying inductive reasoning to some or all of the available evidence. Less confident individuals might hedge their bet on Cournot dynamics with some probabilistic version of the rule. Strategically minded individuals might predict the effect of their current choice on future choices of the other agents involved, and factor this into their decision. Humans, having a very large brain, can do all of these things but often they do not bother (Suppes and Atkinson 1960; Roth and Erev 1995; Erev and Roth 1998; Busemeyer and Stout 2002; Yechiam and Busemeyer 2005).

Reinforcement learning does not require a lot of effort, or a large brain, or any brain at all. In this chapter we will concentrate on reinforcement learning, although we will also touch on other forms of learning. Specifically, we apply a mathematical model in which the probability of an act is proportional to the accumulated rewards from performing that act (Herrnstein 1970; Roth and Erev 1995). Following Luce (1959), the learning model can be decomposed into two parts: (i) a *reinforcement dynamics*, in which weights or propensities for acts evolve, and (ii) a *response rule*, which translates these weights into probabilities of acts. If we let weights evolve by adding the payoff gotten to the weight of the act chosen, and let our probabilities be equal to the normalized weights (Luce's linear response rule), we get the basic Herrnstein–Roth–Erev dynamics.

There are alternative models of reinforcement learning that could be investigated in this setting. In a path-breaking study, Suppes and Atkinson (1960) applied stimulus sampling dynamics to learning in two-person games. Borgers and Sarin (1997) have investigated the dynamics of Bush and Mosteller (1955) in a game-theoretic setting. Instead of the Luce's linear response rule of normalizing the weights, some models use a logistic response rule. Bonacich and Liggett (2003) apply Bush–Mosteller learning in a setting closely resembling our own. They get limiting results that are closely connected to those of the discounted model of Friends II in Skyrms and Pemantle (2000). Liggett and Rolles (2004) generalize the results of Bonacich and Liggett to an infinite space of agents. We, however, will concentrate attention on the basic Herrnstein–Roth–Erev dynamics and on a "slight" variation on it.

Erev and Roth (1998) suggest modifying the basic model by discounting the past to take account of "forgetting". At each time period, accumulated weights are multiplied by some positive discount factor less than one, while new reinforcements are added at full strength. Discounting is a robust phenomenon in experimental studies of reinforcement learning, but there seems to be a great deal of individual variability with reported discount factors ranging from .5 to .99 (Erev and Roth 1998; Busemeyer and Stout 2002; Yechiam and Busemeyer 2005; Goeree and Holt 2002). Discounting changes the limiting properties of the learning process radically. We will see that within the reported range of individual variability, small variations in the discount rate can lead to large differences in predicted observable outcomes in interactive learning situations.

3. Two-Person Games with Basic Reinforcement Learning

We begin by investigating basic (undiscounted) reinforcement learning in simple two-person interactions. The following model was introduced in Skyrms and Pemantle (2000). Each day each individual in a small group wakes up and decides to visit someone. She decides by chance, with the chance of visiting anyone else in the group being given by normalized weights for that individual. (We can imagine the process starting with some initial weights; they can all be set to one to start the process with random encounters.) The person selected always accepts, and there is always time enough in the day for all selected interactions to take place. In a group of ten, if Jane decides to visit someone and the other nine all happen to decide to visit Jane, she has a total of ten interactions in that day. Each interaction produces a payoff. At the end of the day, each individual updates her weights for every other individual by adding the payoffs gotten that day from interactions with that individual. (Obvious variations on the basic model suggest themselves, but we confine ourselves here to just this model applied to different kinds of interactions.) Initially, we investigate baseline cases where individuals have only the choice of with whom to interact, and interactions always produce payoffs in the same way. Then we build on the results for these cases to analyze interactions in the Stag Hunt game, in which different agents can have different acts and the combination of acts determines the payoffs.

Consider two games of "Making Friends." In Friends I the visitor is always treated well, and gains a payoff of 1, while the host goes to some trouble but also enjoys the encounter, for a net payoff of zero. In Friends II the visitor and host are both equally reinforced, with a payoff of 1 going to each. We start each learning process with each individual having initial weights of one for each other individual, so that our group begins by interacting at random. It is easy to run computer simulations of the Friends I and Friends II processes, and it is a striking feature of such simulations that in both cases a non-random interaction structure rapidly emerges. Furthermore, rerunning the processes from the same starting point seems to generate a different structure each time. In this setting, we should expect the emergence of structure without an organizer, or even an explanation in terms of payoff differences. The state of uniform random encounters with which we started the system does not persist, and so must count as a very artificial state. Its use as the fixed interaction structure in many game theoretic models is therefore extremely suspect.

We can understand the behavior of the Friends I process if we notice that each individual's learning process is equivalent to a Pólya urn. We can think of him as having an urn with balls of different colors, one color for each other individual. Initially there is one ball of each color. A ball is chosen (and returned), the designated individual is visited. Because visitors are always reinforced, another ball of the same color is added to the urn. Because only visitors are reinforced, balls are not added to the urn in any other way. (Philosophers of science will be familiar with the Pólya urn because of its equivalence with Bayes–Laplace inductive inference.) The Pólya urn converges to a limit with probability one, but it is a random limit with uniform distribution over possible final probabilities.

Anything can happen, and nothing is favored! In Friends I the random limit is uniform for each player, and makes the players independent (Skyrms and Pemantle 2000: Theorem 1). All interaction structures are possible in the limit, and the probability that the group converges to random encounters is zero.

In Friends II, both visitor and host are reinforced and so the urns interact. If someone visits you, you are reinforced to visit him—or to put it graphically, someone can walk up to your door and put a ball of his color in your urn. This complicates the analysis. Nevertheless, the final picture is quite similar. The limiting probabilities must be *symmetric*, that is to say X visits Y with the same probability that Y visits X, but subject to this constraint and its consequences anything can happen (Skyrms and Pemantle 2000: Theorem 2).

So far, the theory has explained the surprising results of the simulations, but a rather special case of Friends II provides a cautionary contrast. Suppose that there are only three individuals. (What we are about to describe is much less likely to happen if the number of individuals is a little larger.) Then the only way we can have symmetric visiting probabilities is if each individual visits the other two each with probability one-half. Then the previous theorem implies that in this case the process must converge to these probabilities. In simulations this sometimes happens rapidly. However, there are other trials in which the system appears to be converging to a state in which individual A visits B and C equally, but B and C always visit A and never each other. You can think of individual A as "Ms. Popular." The system was observed to stay near such a state for a long time (5,000,000 iterations of the process).

This apparent contradiction is resolved in Pemantle and Skyrms (2004), using the theory of stochastic approximation. For the basic Herrnstein–Roth–Erev model, there is an underlying deterministic dynamics that can be obtained from the expected increments of the stochastic process. This deterministic dynamics has four equilibria—one in which each individual visits the others with equal probability and the other three having A, B, and C respectively as "Ms. Popular." The symmetric equilibrium is strongly stable—an attractor—while the "Ms. Popular" equilibria are unstable saddle points. The system must converge to the symmetric equilibrium. It cannot converge to one of the unstable saddles, but if in the initial stages of learning it wanders near a saddle it may take a long time to escape because the vector pushing it away is very small. This is what happens in the anomalous simulations. There is a methodological moral here that we will revisit in the next section. Simulations may not be a reliable guide to limiting behavior and limiting behavior is not necessarily all that is of interest.

The Making Friends games provide building blocks for analyzing learning dynamics where the interactions are games with non-trivial strategies. Consider the two-person Stag Hunt. Individuals are either Stag Hunters or Hare Hunters. If a Stag Hunter interacts with a Hare Hunter no Stag is caught and the Stag Hunter gets zero payoff. If a Stag Hunter interacts with another Stag Hunter the Stag is likely caught and the hunters each get a payoff of one. Hare Hunting requires no cooperation, and its practitioners get a payoff of .75 in any case. The game is of special interest for social theory because cooperation is both mutually beneficial and an equilibrium, but it is risky (Skyrms 2004). In game theoretic terminology, Stag hunting is payoff dominant and Hare hunting is risk dominant. In a large population composed of half Stag Hunters and half Hare Hunters with random interactions between individuals, the Hare Hunters would get an average payoff of .75 while the Stag Hunters would only get an average payoff of .50. The conventional wisdom is that in the long run evolution will strongly favor Hare hunting, but we say that one should consider the possibility that the players *learn to network*.

We use exactly the same model as before, except that the payoffs are now determined by the individuals' types or strategies: Hunt Stag or Hunt Hare. We start with an even number of Stag Hunters and Hare Hunters. Theory predicts that, in the limit, Stag Hunters always visit Stag Hunters and Hare Hunters always visit Hare Hunters (Skyrms and Pemantle 2000: Th. 6). Simulation confirms that such a state is approached rapidly. Although on rational choice grounds Hare Hunters "should not care" whom they visit, they cease to be reinforced by visits from Stag Hunters after Stag Hunters learn not to visit them. Hare Hunters continue to be visited by other Hare Hunters, so all the differential learning for Hare Hunters takes place when they are hosts rather than visitors. Once learning has sorted out Stag Hunters and Hare Hunters so that each group only interacts with its own members, each is playing Friends II with itself and previous results characterize within-group interaction structure.

Now Stag Hunters prosper. Was it implausible to think that Stag Hunters might find a way to get together? If they were sophisticated, well-informed, optimizing agents they would have gotten together right away! Our point is that it doesn't take much for Stag Hunters to get together. A little bit of reinforcement learning is enough.

4. Clique Formation with Discounting the Past

Adding a little discounting of the past is a natural and seemingly modest modification of the reinforcement process. However, it drastically alters the limiting behavior of learning. If the Pólya urn, which we used in the analysis of Friends I, is modified by discounting the past, the limiting result is that after some time (can't say when) there will be one color (can't say which) that will always be picked. Discounting the past, no matter how little the discounting, leads to deterministic outcomes. This is also true when we learn to network. Discounting the past leads to the formation of cliques, whose members never interact with members of alternative cliques. Why then, did we even bother to study learning without discounting? We will see that if discounting is small enough, learning with discounting may, for long periods of time, behave like learning without discounting.

The effects of adding discounting to the learning process are already apparent in two-person interactions (Skyrms and Pemantle 2000), but they are more interesting in multi-person interactions. Here we discuss two three-person interactions, Three's Company (a uniform reinforcement counterpart to Friends II), and a Three-Person version of the Stag Hunt. Every day, each individual picks two other individuals to visit to have a three-person interaction. The probability of picking a pair of individuals is taken to be proportional to the product of their weights. The payoff that an individual receives from a three-person interaction is added to her weights for each of the other two participants. We again start the learning process with random interaction. Everyone begins having weight one for everyone else.

In Three's Company, as in Friends II, everyone is always reinforced in every interaction. Everyone gets a payoff of one. No matter what the discount rate, the limiting result of discounted learning is clique formation. For a population size of six or more, the population will break up into cliques of size 3, 4, or 5. Each member of a given clique chooses each other member of that clique with positive limiting relative frequency. For each member of a clique, there is a finite time after which she does not choose outsiders. All such cliques—that is each partition of the population into sets of size 3, 4, and 5, has positive probability of occurring (Pemantle and Skyrms 2004: Th. 4.1).

Simulations at a discount rate of .5 conform to theory. A population of 6 always broke into two cliques of size 3, with no interactions between cliques. As we discount less—keeping more of the past weights—we see a rapid shift in results. Multiplying past weights by .6, led to the formation of 2 cliques in 994/1000 trials; by .7 in 13/1000; by .8 in none. (We ran the process for 1,000,000 time steps and rounded interaction probabilities to two decimal places.) Writing the discount factor by which past payoffs are multiplied as $(1-x)$, we can say that simulation says that clique formation occurs reliably for large x, but not at all for small x, with a large transition taking place between $x = .4$ and $x = .3$. The theory says that clique formation occurs for any positive x.

This apparent conflict between theory and simulation is resolved in Pemantle and Skyrms (2003), where it is shown that time to clique formation increases exponentially in $1/x$ as the discount factor $(1-x)$ approaches 1. The behavior of the process for observable finite sequences of iterations is highly sensitive to the discount parameter, within ranges that fall within the individual variability that has been reported in the experimental literature. When x is close to 1, discounted reinforcement learning behaves for long periods of time like undiscounted learning in which clique formation almost never occurs.

Three's Company, like Friends II, is important because it arises naturally in the analysis of less trivial interactions. Consider a Three-Player Stag Hunt (Pemantle and Skyrms 2003). Pairs of individuals are chosen, and weights evolve, just as in Three's Company, but the payoffs depend on the types of players. If three Stag Hunters interact, they all get a payoff of 4, but a Stag Hunter who has at least one Hare Hunter in his trio gets nothing. (In a random encounter setting, Stag Hunting is here even more risky that in the two-person case.) Hare hunters always get a payoff of 3.

In the limit Stag Hunters learn to always visit other Stag Hunters but, unlike some other limiting results we have discussed, this one is attained very rapidly. With 6 Stag Hunters and 6 Hare Hunters and a discount rate of .5, the probability that a Stag Hunter will visit a Hare Hunter usually drops below half a percent in 25 interactions. In 50 iterations this always happened in 1000 trials, and this remains true for values of x between .5 and .1. For $x = .01$, 100 iterations suffices and 200 iterations are enough if $x = .001$.

Once Stag Hunters learn to visit Stag Hunters, they are essentially playing a game of Three's Company among themselves. They may be visited by Hare Hunters, but these visits produce no reinforcement for the Stag Hunters and so do not alter their weights. Stag Hunters then form cliques of size 3, 4, or 5 among themselves. This will take a long time if the past is only slightly discounted.

There is a tendency for Hare Hunters to learn to visit Hare Hunters after the Stag Hunters learn not to visit them, but because of the discounting it is possible for a Hare Hunter to be frozen in a state of visiting one or two Stag Hunters. This is a real possibility when the past is heavily discounted. At $x = .5$, at least one Hare Hunter interacted with a Stag Hunter (after 10,000 iterations) in 384 out of 1000 trials. This dropped to 6/1000 for $x = .2$ and to 0 for $x = .1$. Hare Hunters who are not trapped into interactions with Stag Hunters eventually end up playing Three's Company among themselves and also form cliques of size 3, 4, and 5.

5. Coevolution of Structure and Strategy

So far we have concentrated on the dynamics of interaction, because we believe that it has not received as much attention as it deserves. The full story involves co-evolution of both interaction structure and strategy. Depending on the application, these may involve the same or different adaptive dynamics and they may evolve at the same or different rates. We will illustrate this with two different treatments of the two-person Stag Hunt.

To the two-person Stag Hunt of Section 3, we add a strategy revision process based on imitation. This *reinforcement-imitation* model was discussed in Skyrms and Pemantle (2000). With some specified probability, an individual wakes up, looks around the whole group, and if some strategy is prospering more than his own, switches to it. Individual's probabilities are independent. If imitation is fast relative to structure dynamics, it operates while individuals interact more or less at random and Hare Hunters will take over more often than not. If imitation is slow, Stag hunters find each other and prosper, and then imitation slowly converts Hare Hunters to Stag Hunters (who quickly learn to interact with other Stag Hunters).

Simulations show that in intermediate cases, timing can make all the difference. We start with structure weights equal to 1 and vary the relative rates of the dynamics by varying the imitation probability. With "fast" imitation (pr = .1) 78 percent of the trials ended up with everyone converted to Hare Hunting and 22 percent ended up with everyone converted to Stag Hunting. Slower imitation (pr = .01) almost reversed the numbers, with 71 percent of the trials ending up All Stag and 29 percent ending up All Hare. Fluid network structure coupled with slow strategy revision reverses the orthodox prediction that Hare Hunting (the risk dominant equilibrium) will take over.

(This conclusion remains unaffected if we add discounting to the learning dynamics for interaction structure. Discounting the past simply means that Stag Hunters find each other more rapidly. No matter how Hare Hunters end up, Stag Hunters are more prosperous. Imitation converts Hare Hunters to Stag Hunters.)

The foregoing model illustrates the combined action of two different dynamics, reinforcement learning for interaction structure and imitation for strategy revision. *What happens if both processes are driven by reinforcement learning?* In particular, we would like to know whether the relative rates of structure and strategy dynamics still make the same difference between Stag Hunting and Hare Hunting. In this *Double Reinforcement* model, each individual has two weight vectors, one for interaction propensities and one for propensities to either Hunt Stag or Hunt Hare. Probabilities for whom to visit and what to do are both obtained by normalizing the appropriate weights. Weights are updated by adding the payoff from an interaction to both the weight for the individual involved and the weight for the action taken. Relative rates of the two learning processes can be manipulated by changing the magnitude of the initial weights.

In the previous models we started the population off with some Stag Hunters and some Hare Hunters. That point of view is no longer correct. The only way one could be deterministically a Stag Hunter would be if he started out with zero weight for Hare Hunting, and then he could never learn to hunt Stag. We have to start out individuals with varying propensities to hunt Hare and Stag. There are various interesting choices that might be made here; we will report some simulation results for one. We start with a group of 10, with 2 confirmed Stag Hunters (weight 100 for Stag, 1 for Hare), 2 confirmed Hare Hunters (weight 100 for Hare,

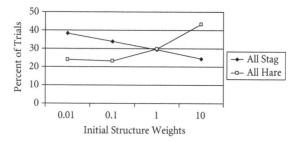

Figure 10.1 Stag Hunt with Reinforcement Dynamics for both Strategy and Structure

1 for Stag), and 6 undecided guys (weights 1 for Stag and 1 for Hare. Initial weights for interaction structure were all equal, but their magnitude was varied from .001 to 10, in order to vary the relative rates of learning structure and strategy. The percentage of 10,000 trials that ended up All Stag or All Hare (after 1,000,000 iterations) for these various settings are shown in Figure 10.1. As before, fluid interaction structure and slow strategy adaptation favor Stag Hunting, while the reverse combination favors Hare Hunting.

In both reinforcement-imitation and double reinforcement models of the co-evolution of structure and strategy a fluid network structure shifts the balance from the risk dominant Hare Hunting equilibrium to the cooperative Stag Hunt.

6. Why Dynamics?

Classical, pre-dynamic, game theory would approach the problem differently. The whole group of 10 individuals is playing a 10-person game. A move consists in choosing both a person to play with and a strategy. We can just identify the Nash equilibria of this large game. None are strict. The pure equilibria fall into two classes. One class has everyone hunting Stag and every possible interaction structure. The other has everyone hunting Hare and every possible interaction structure. (There are also mixed equilibria with every possible interaction structure.) From this point of view, interaction structure does not seem very important. If you ignore dynamics you miss a lot.

References

Alexander, J. M. (2000) "Evolutionary Explanations of Distributive Justice." *Philosophy of Science* 67: 490–516.

Bala, V. and S. Goyal (2000) "A Non-Cooperative Model of Network Formation." *Econometrica* 68: 1181–229.

Bonacich, P. and T. Liggett (2003) "Asymptotics of a Matrix-Valued Markov Chain Arising in Sociology." *Stochastic Processes and Their Applications* 104: 155–71.

Borgers, T. and R. Sarin (1997) "Learning Through Reinforcement and Replicator Dynamics." *Journal of Economic Theory* 77: 1–14.

Busemeyer, J. and J. Stout (2002) "A Contribution of Cognitive Decision Models to Clinical Assessment: Decomposing Performance on the Bechara Gambling Task." *Psychological Assessment* 14: 253–62.

Bush, R. and F. Mosteller (1955) *Stochastic Models of Learning*. New York: John Wiley & Sons.

Ellison, G. (1993) "Learning, Local Interaction, and Coordination." *Econometrica* 61: 1047–71.

Erev, I. and A. Roth (1998) "Predicting How People Play Games: Reinforcement Learning in Experimental Games with Unique Mixed Strategy Equilibria." *American Economic Review* 88: 848–81.

Goeree, J. K. and C. A. Holt (2002) "Learning in Economic Experiments." In *Encyclopedia of Cognitive Science*, Volume 2, ed. L. Nagel, 1060–9. Macmillan: New York.

Hegselmann, R. (1996) "Social Dilemmas in Lineland and Flatland." In *Frontiers in Social Dilemmas Research*, ed. W. Liebrand and D. Messick, 337–62. Berlin: Springer.

Herrnstein, R. J. (1970) "On the Law of Effect." *Journal of the Experimental Analysis of Behavior* 13: 243–66.

Jackson, M. and A. Watts (2002) "On the Formation of Interaction Networks in Social Coordination Games." *Games and Economic Behavior* 41: 265–91.

Kirman, A. (1997) "The Economy as an Evolving Network." *Journal of Evolutionary Economics* 7: 339–53.

Liggett, T. M. and S. Rolles (2004) "An Infinite Stochastic Model of Social Network Formation." *Stochastic Processes and Their Applications* 113: 65–80.

Luce, R. D. (1959) *Individual Choice Behavior*. New York: John Wiley and Sons.

Nowak, M. and R. May (1992) "Evolutionary Games and Spatial Chaos." *Nature* 359: 826–9.

Pemantle, R. and B. Skyrms (2003) "Time to Absorption in Discounted Reinforcement Models." *Stochastic Processes and Their Applications* 109: 1–12.

Pemantle, R. and B. Skyrms (2004) "Network Formation by Reinforcement Learning: The Long and the Medium Run." *Mathematical Behavioral Sciences* 48: 315–27.

Roth, A. and I. Erev (1995) "Learning in Extensive Form Games: Experimental Models and Simple Dynamic Models in the Intermediate Term." *Games and Economic Behavior* 8: 14–212.

Skyrms, B. (2004) *The Stag Hunt and the Evolution of Social Structure.* New York: Cambridge University Press.

Skyrms, B. and R. Pemantle (2000) "A Dynamic Model of Social Network Formation." *Proceedings of the National Academy of Sciences of the USA* 97: 9340–6.

Suppes, P. and R. Atkinson (1960) *Markov Learning Models for Multiperson Interactions.* Palo Alto: Stanford University Press.

Yechiam, E. and J. R Busemeyer (2005) "Comparison of Basic Assumptions Embedded in Learning Models for Experienced Based Decision-Making." *Psychonomic Bulletin and Review* 12: 387–402.

11

A Dynamic Model of Social Network Formation

with Robin Pemantle

Pairs from among a population of ten individuals interact repeatedly. Perhaps they are cooperating to hunt stags and rabbits, or coordinating on which concert to attend together; perhaps they are involved in the somewhat more antagonistic situation of bargaining to split a fixed payoff, or attempting to escape the undesirable but compelling equilibrium of a Prisoner's Dilemma. As time progresses, the players adapt their strategies, perhaps incorporating randomness in their decision rules, to suit their environment. But they may also exert control over their environment. The players may have choice over the pairings but not perfect information about the other players. They may improve their lot in two different ways. A child who is being bullied learns either to fight better or to run away. Similarly, a player who obtains unsatisfactory results may choose either to change strategies or to change associates. Regardless of whether the interactions are mostly cooperative or mostly antagonistic, it is natural and desirable to allow evolution of the social network (the propensity for each pair to interact) as well as the individuals' strategies.

We build a model that incorporates both of these modes of evolution. The idea is simple:

(*) Individual agents begin to interact at random. The interactions are modeled as games. The game payoffs determine which interactions are reinforced, and the social network structure emerges as a consequence of the dynamics of the agents' learning behavior.

As the details of the specific game and the reinforcement dynamics vary, we then obtain a class of models. In this chapter, we treat some simple reinforcement dynamics, which may serve as a base for future investigation.

The idea of the simultaneous evolution of strategy and social network appears to be almost completely unexplored. Indeed, the most thoroughly studied models of evolutionary game theory assume *mean-field* interactions, where each individual is always equally likely to interact with each other. Standard treatments of evolutionary game dynamics (Hofbauer and Sigmund 1988; Weibull 1997) operate entirely in this paradigm. This is due, to a large extent, to considerations of theoretical tractability of the model. Models have been introduced that allow the agents some control over their choice of partner (Feldman and Thomas 1987), but the control is still exerted in a mean-field setting: one chooses between the present partner and a new pick at random from the whole population.

Evolutionary biologists know that evolutionary dynamics can be affected by nonrandom encounters or population structure, as in Sewall Wright's (1921) models of assortative mating. Wright (1945) realized that a positive correlation of encounters could provide an account of the evolution of altruism. Thus, the need for social network models has been long recognized.

When the social network is modeled, it is almost always static.[1] Interactions, for example, may be posited to occur only between players whose locations are close, according to some given spatial data. Biological models in which encounters are governed by spatial structure have become increasingly frequent in the 1990s; see, for example, the work of Durrett, Levin, and Neuhauser (Durrett and Levin 1994; Kang, Krone, and Neuhauser 1995; Durrett and Neuhauser 1997). A similar hypothesis of spatial structure, in a game theory context, arises in Blume (1993). Here, technology from statistical mechanics is adapted to the analysis of games whose interactions take place between neighbors in a grid.

A number of recent investigations by game theorists, some directly inspired by biological models, have shown that the dynamics of strategic

---

[1] An exception, perhaps, is a preprint we have recently encountered by Jackson and Watts (1999).

interaction can be strikingly different if interaction is governed by some spatial structure, or more generally, some graph structure (Pollack 1989; Lindgren and Nordahl 1994; Anderlini and Ianni 1997). For instance, one-shot Prisoner's Dilemma games played with neighbors on a circle or torus allows cooperation to evolve in a way that the random encounter model does not. The spatial or graph structure can be important to determine which equilibria are possible, whether repeated interactions can be expected to converge to equilibrium, and, if so, how quickly convergence takes place (Ellison 1993).

Because the outcome of a repeated game may vary with the choice of network model, it is important to get the network model right. Further progress in the theory of games and adaptive strategies would be greatly enhanced by a theory of networks of social interaction. In particular, it would be desirable to have a framework within which models may be developed that are both tractable and plausible as a mechanism governing interactions among a population of agents seeking to improve their lot.

When the network changes much more slowly than do the strategies of individuals, it is reasonable to model the social network by a structure that is fixed, though possibly random. The question of realistically modeling the randomness in such a case is taken up in a number of papers, of which a recent and well-known example is the "small world" model (Watts and Strogatz 1998). In the other extreme (Schelling 1969, 1971; Epstein and Axtell 1996), the evolution of social structure is modeled by agents moving on a fixed graph in the absence of strategy dynamics.

In the general case, however, interaction structures are fluid and evolve in tandem with strategy. What is required here is a dynamics of interaction structure to model how social networks are formed and modified. We distinguish this *structure dynamics* from the *strategic dynamics* by which individuals change their individual behaviors or strategies.

In this chapter, we introduce a simple, additive model for structure dynamics, and we explore the resulting system under several conditions: with or without discounting of the past, with or without added noise, and in the presence or absence of strategic dynamics. Common to all our models is a stochastic evolution from a (usually symmetric) initial state. Individuals in a population start out choosing whom to interact with at random and then modify their choices according to how their choice is reinforced, and then the process is repeated. An infinite variety of such models is possible. We will consider only a few basic models, meant to illustrate that rigorous results on structure dynamics are not out of reach, and that further inquiry will be profitable.

We first consider a baseline case of uniform reinforcement. Here, any choice of partner is reinforced as strongly as any alternative choice would have been. In other words, the interaction game between any pair of players always produces a constant reward or punishment. One might expect that such cases would not lead to interesting dynamics, but that is far from the truth. We show both by computer simulation and analytically how structure emerges spontaneously even in these cases. Because the strategic dynamics here are trivial, the baseline case is intended mostly as a building block on which more interesting strategic dynamics are to be grafted. We note, however, that the constant reward game is not completely unreasonable. Studies have shown that in the absence of other environmental attributes, sheer familiarity brings about positive attitudinal change (Zajonc 1968). In fact, an abstract model of network evolution under uniform positive reweighting has appeared before under the name of "Reinforced Random Walk" (D. Coppersmith and P. Diaconis, unpublished work).

Next, we move to the case where players of different types play a nontrivial game and are reinforced by the payoffs of the game. Here, we examine the coevolution of behavior and structure when the structural dynamics and strategic dynamics are both operative. The relative speeds of structural dynamics and strategic dynamics affect which equilibrium is selected in the game. In particular, this can determine whether the risk-dominant or payoff-dominant equilibrium is selected.

1. Making Friends: A Baseline Model of Uniform Reinforcement

Friends I: Asymmetric weights

Each morning, each agent goes out to visit some other agent. The choice of whom to visit is made by chance, with the chances being determined by the relative *weights* each agent has assigned to the others. For this purpose, agent number i has a vector of weights $<w_{i1}, \ldots, w_{in}>$ that she assigns to other players (assume $w_{ii} = 0$). Then she visits agent j with probability

$$\text{Prob}(\text{agent } i \text{ visits } j) = \frac{w_{ij}}{\sum_k w_{ik}}. \tag{1.1}$$

Here we are interested in a symmetric baseline model, so we will assume that all initial weights are 1. Initially, for all agents, all possible visits are equiprobable.

Every agent is treated nicely on her visit and all are treated equally nicely. They each get a reinforcement of 1. Each agent then updates her weight vector by adding 1 to the weight associated with the agent that she visited. Her probabilities for the next round of visits are modified accordingly. At each stage, we have a matrix p_{ij} of probabilities for i to visit j. Do these probabilities converge, and if so to what?

Given all the symmetry built into the starting point and the reinforcement, it is perhaps surprising that all sorts of structures emerge. Here is a description of a simulated sample run of length 1,000. The probabilities, to two decimal places, seem to converge after a few hundred rounds of visits, to a matrix that is anything but uniform (and to a different matrix each time the process is run from the initial, symmetric weights). There is one agent, A, who visits another agent, B, more than half the time. There is no reciprocation, so this has no bearing on how often B visits A, and in fact most agents will not visit any one agent more than a third of the time.

In the analysis section, we show that this outcome is typical.

THEOREM 1. *The probability matrix for Friends I with n players will converge to a random limit p as time goes to infinity. The distribution of the limit is that the rows of p are independent, each having Dirichlet distribution (ignoring the zero entry on the diagonal) whose parameters are $n - 1$ ones.*

Thus we see a spontaneous emergence of structure. This type of simple model has been used before in the economics literature to explain the stabilization of market shares at seemingly random equilibria, due to random reinforcement in the early phases of growth of an industry (Arthur 1989). We remark that the choices made by each agent are independent of the choices made by each other agent, so the social aspect of the model is somewhat degenerate and the model may be viewed as a model of individual choice. Nevertheless, it fits our definition of social network model in that it gives a probabilistic structure to interactions; one may then extend the model so the interactions are nontrivial games.

Friends II: Symmetrized reinforcement

Suppose now that the interaction is as pleasant to the host as the visitor. Thus when agent i visits agent j, we add 1 to both w_{ij} and w_{ji}. A typical outcome for ten agents after 1,000 rounds of visits looks similar to the table for Friends I, except that the entries are nearly symmetric. There are, however, subtle differences that may cause the two models to act very differently when strategic dynamics are introduced. To see these differences, we describe what is typically observed after ten runs of a simulation of Friends II to time 1,000 for a set of three agents, this being the minimum population size for which structural dynamics are interesting. What we see typically is one or two runs in which each player's visits are split evenly (to two decimal places) between the others. We see another several runs that are close to this. We see one run or so in which two agents nearly always visit the third agent, which splits its time among the other two. The remaining runs give something between these extreme outcomes.

What may not be apparent from such data is that the limiting weights for Friends II are always 1/2. Only a small fraction of sample outcomes decisively exhibit the proven limiting behavior. The data, in other words, show that after 1,000 iterations, the weights may still be far from their limiting values; when this is the case, one of the three agents is largely ignored by the other two and visits each of the other two herself equally often. Because the lifetime of many adaptive games is 1,000 rounds or fewer, we see that limiting behavior may not be a good guide to behavior of the system on the time scales we are interested in. The analysis section discusses both limiting results for this model and finite time behavior. When the population size is more than 3, the weights will always converge, but the limit is random and restricted to the subspace of symmetric matrices. Again, convergence of the weights to their limiting values is slower than in the nonreciprocal game of Friends I.

THEOREM 2. *The probability matrix p_{ij} for Friends II with n players converges to random limit p as time goes to infinity. If n = 3, the limit is the matrix all of whose off-diagonal entries are 1/2. In general, the limit may be any symmetric matrix whose rows sum to 1; that is, the closed support of the random limit is the entire subspace of symmetric stochastic matrices.*

Analysis of Friends I and II

To fit this in the framework of (*), construct the following degenerate games. Each of the two players has only one strategy, and the payoff matrix is as follows.

Friends I	Host	Friends II	Host
Visitor	(1, 0)	Visitor	(1, 1)

The weights w_{ij} are initialized to 1 for $i \neq j$, and are then updated according to

$$w_{ij}(t + 1) = w_{ij}(t) + u(i, j; t), \qquad (1.2)$$

where $w_{ij}(t)$ is the weight agent i gives to agent j at time t and $u(i, j; t)$ is the payoff of the game played at time t between visitor i and host j (and zero if this visit did not occur at time t). This, together with specification of the visitation probabilities in Eq. (1.1), defines the model. Changing the initial weights does not affect the qualitative behavior of any model, so there is no need to vary the initialization.

For Friends I, the updating of the weights for any one agent is the same as a Pólya urn process (Eggenberger and Pólya 1923). Each agent can be thought of as having an urn with balls of $n - 1$ colors, one color representing each other agent. Initially there is one ball of each color in the urn. The agent picks a ball at random, indicating whom she should visit, then returns it to the urn along with an extra ball of the same color. The urns belonging to different agents are statistically independent.

The analysis of this process is well-known (Johnson and Kotz 1977: ch. 4). It is easy to show that the sequence of draws for each agent is *exchangeable*, that is, permuting a sequence does not change its probability. Hence by the de Finetti representation theorem, the random sequence of draws from an urn is equivalent to a mixture of multinomial processes, that is, of sequences of independent draws. The mixing measure is easily seen to be Dirichlet. Consequently, the visiting probabilities converge with probability one, but they can converge to anything. That they converge to the uniform vector, where each agent has equal probability to visit each other, has prior probability zero.

Furthermore, convergence to the limiting probability matrix is quite rapid. Let $p(t)$ denote the matrix whose (i, j)-entry is $p_{ij}(t)$. Then exchangeability implies that, conditional on the limit matrix $p = \lim_{t \to \infty} p(t)$, the sequence of visits is a sequence of independent, identically distributed draws from the limit distribution. Thus at time t, the central limit theorem implies that $p(t) - p$ is $t^{-1/2}$ times a multivariate normal.

For Friends II, exchangeability fails. This is not surprising, because the property of exchangeability is not very robust. More surprising, however, is that the sequence of probability matrices $p(t)$ does not form a martingale. To explain this terminology, let E_t denote the expectation conditioned on the values at time t. A simple computation shows that for Friends I, the expected value of $p_{ij}(t+1)$ conditioned on the time t value is equal to $p_{ij}(t)$: because w_{ij} increases only when i visits j, we have

$$
\begin{aligned}
E_t p_{ij}(t+1) &= E_t \sum_{k=1}^{n} p_{ik}(t) \frac{w_{ij} + \delta_{jk}}{1 + \sum_{l=1}^{n} w_{il}(t)} \\
&= \frac{w_{ij}(t) + p_{ij}(t)}{1 + \sum_{l=1}^{n} w_{il}(t)} \\
&= p_{ij}(t).
\end{aligned}
$$

Even without exchangeability, the martingale convergence theorem (Durrett 1996: sect. 4.2) implies convergence of the quantities p_{ij}, though it says very little about the limit.

For Friends II, a complete analysis may be obtained (R.P. and B.S., unpublished work). Here is an outline of what is found there. A computation similar to the one for Friends I shows that

$$
E_t p(t+1) = p(t) + \frac{1}{t} F(p(t)),
$$

where F is a certain function on symmetric n by n matrices. In other words, the random sequence of matrices $\{p(t) : t = 1, 2, \dots\}$ is a *stochastic approximation* in the sense of Robbins and Monro (1951), driven by the vector field F. General results of Pemantle (1990) and Benaïm and Hirsch (1995) now imply that $p(t)$ converges to the set where F vanishes. To show that $p(t)$ always converges to a single point, Pemantle and Skyrms (unpublished work) compute a Liapunov function for F, that is, a function V for which $\nabla V \cdot F < 0$ with equality only when $F = 0$. This, together with an efficiency inequality (bounding the angle between f and ∇V away from ninety degrees), establish convergence of p. The remainder of *Theorem 2* is then established by showing the only stable zeros of the vector field F are the symmetric matrices with row sums all equal to 1, and that the possible limit points of $p(t)$ are exactly the stable equilibria of the flow determined by F.

Determination of the rate of convergence of $p(t)$ to its limit is somewhat different in this case. Because of the presence of unstable equilibria from the flow determined by F, there is a possibility of being stuck near one of these equilibria for a long time before eventually following the flow to one of the stable equilibria. For the three-player game, the unstable equilibria are the following three matrices:

$$
\begin{pmatrix} 0 & \frac{1}{2} & \frac{1}{2} \\ 1 & 0 & 0 \\ 1 & 0 & 0 \end{pmatrix}
\begin{pmatrix} 0 & 1 & 0 \\ \frac{1}{2} & 0 & \frac{1}{2} \\ 0 & 1 & 0 \end{pmatrix}
\begin{pmatrix} 0 & 0 & 1 \\ 0 & 0 & 1 \\ \frac{1}{2} & \frac{1}{2} & 0 \end{pmatrix}.
$$

These correspond to cases where one of the three agents is always visited and splits her visits equally between the other two. The probability that $p(t)$ is within ϵ of one of these traps is roughly $3\epsilon t^{-1/3}$, so with $t = 1,000$, we find a reasonably high probability that $p(1,000)$ is not near the uniform probability matrix but is instead still near one of the unstable equilibria. This persists with reasonable probability well beyond $t = 10^6$. For greater population sizes, similar phenomena apply. Convergence to the invariant set is relatively slow. However, for large populations, say twenty or more, another phenomenon takes place. The portion of the space of possible p matrices that are within ϵ of the possible limits goes to 1; this is known as the concentration of measure phenomenon (Talagrand 1995). Thus it becomes very unlikely to get stuck initially far away from the limit, simply because the initial randomness will very likely lead to a point very near a possible limit. Thus for large populations, the dynamics appear very similar to the dynamics for Friends I.

2. Making Enemies

Let us change the "Making Friends" model in just one way. Instead of being rewarded, agents are punished; instead of uniformly positive interactions, we have uniformly negative ones:

Enemies I	Host	Enemies II	Host
Visitor	$(-1, 0)$	Visitor	$(-1, -1)$

Instead of interactions being reinforcing, we take them as inhibiting. The dynamics of inhibition might be modeled in a number of ways. Continuing to use the update Eq. (1.2) will not work because the weights will end up becoming negative and the visitation probabilities in Eq. (1.1) will be meaningless. In this section, we explore two other possible rules for updating the weights so as to inhibit past behavior. With negative reinforcement, it is easy to predict what will happen: the social network always becomes uniform, and the dynamics are not sensitive to the particular updating mechanism. Indeed, this is what happens. Because there are no surprises, and because this model is just a building block for a model with both structural and strategic dynamics, we keep the discussion brief.

The transfer model

Consider a three-player model with the following update rule on the weights. Initial weights are all positive integers. When i visits j, the weight w_{ij} is diminished by 1 and the weight w_{ik}, $k \neq i, j$, is increased by 1. This is equivalent to the Ehrenfest model of heat exchange between two bodies (Ehrenfest and Ehrenfest 1907). In the original Ehrenfest model, there are two urns. A ball is drawn at random from among all balls in both urns and transferred to the other urn. The distribution of balls tends to the binomial distribution, where each ball is independently equally likely to be in either urn. In Making Enemies, with transfer dynamics and three players, each player may be thought of as having such a pair of urns. The urns are independent.

Because the number of balls is fixed, an Ehrenfest urn is a Markov chain with a finite number of states, where the states consist of distributions over the two urns. For example, if there are only two balls, then there are three states, $S1$, $S2$ and $S3$, corresponding to urn cardinalities of $(2, 0)$, $(1, 1)$, and $(0, 2)$. The transition matrix for this Markov chain is

$$\begin{pmatrix} 0 & 1 & 0 \\ \frac{1}{2} & 0 & \frac{1}{2} \\ 0 & 1 & 0 \end{pmatrix},$$

and the unique stationary vector is $(1/4, 1/2, 1/4)$. In contrast to the Pólya urn, we do not have convergence of the conditional probabilities of visits at each stage given the present: at any time, given the present composition, the probability of a given visit may be 0, 1/2, or 1, depending on the composition of the urns belonging to the visitor. However, if the number of balls, N, is large, approximately equal visiting probabilities are very likely in the following sense. The invariant distribution is binomial, which is concentrated around nearly even distributions when the number of balls is large. Thus, with high probability, no matter what the initial state, after roughly $N \log N/2$ steps (Diaconis and Stroock 1991), the composition of an urn with N balls will be close to a draw from a binomial distribution. The conditional probability of either of the two possible visits will therefore be close to 1/2 and will tend to remain there with high probability. Kac (1947) uses these properties to resolve the apparent paradoxes that beset Bolzmann's discussion of irreversibility in statistical mechanics.

The resistance model

The transfer model allows for a finite cumulative amount of negative reinforcement, and indeed yields a finite Markov chain. Let us explore a rather different model, termed the *resistance model*, in which negative payoffs generate resistance. Initially every choice has resistance 1. The magnitude of a negative payoff is added to its associated resistance, so Eq. (1.2) becomes

$$w_{ij}(t+1) = w_{ij}(t) + |u(i,j;t)|.$$

In the case at hand, when all payoffs are negative, the probability of i visiting j is proportional to the reciprocal of the resistance:

$$p_{ij} = \text{Prob}(\text{agent } i \text{ visits } j) = \frac{1/w_{ij}}{\sum_{k=1}^{n} 1/w_{ik}}$$

with $1/w_{ii} = 0$ by convention. The dynamics of Enemies I and Enemies II under resistance dynamics are easy to describe.

THEOREM 3. *For Enemies I or Enemies II, from any initial conditions, the probability matrix $p(t)$ converges to the uniform probability matrix $\overline{p}$ where $\overline{p}_{ij} = 1/(n-1)$ for any $i \neq j$. The rate of convergence is rapid: of order $N \log N$ if the initial resistances are of order N. The deviations from uniform obey a central limit theorem:*

$$t^{1/2}(p - \overline{p}) \to X$$

where X is a multivariate normal with covariance matrix of rank n(n − 1) in Enemies I and n(n − 1)/2 in Enemies II. In other words, deviations from uniformity are independent normals, subject to the constraints of adding up to zero for each individual and, in the case of Enemies II, the constraints of symmetry.

The central limit theorem may be derived from a stronger, functional central limit theorem, linearizing the system near the uniform probability to see that the paths

$$t \to N^{-1/2}(p(Nt) - \bar{p})$$

converge in distribution as $N \to \infty$ to a multivariate Ornstein–Uhlenbeck process. The rate of convergence follows from standard coupling arguments.

While uniform positive reinforcement breeds structure from unstructured initial conditions, uniform negative reinforcement evidently breeds uniformity even from structured initial conditions. It would appear, therefore, that the customary random encounter (mean-field) model is more suitable for Making Enemies than Making Friends.

A better model?

We would like a model that allows for both positive and negative reinforcement. A natural choice is to let w_{ij} keep track of the log-likelihood for i to visit j, so that probability of i visiting j is given by

$$p_{ij} = \text{Prob(agent } i \text{ visits } j)$$
$$= \frac{exp(w_{ij})}{\sum_{k=1}^{n} exp(w_{ik})}. \tag{2.1}$$

In the next section, we will see a property this rule has in common with rules that discount the past, namely that it leads to being trapped in a deterministic state where i always visits the same j.

QUESTION 1. *Is there a model incorporating both positive and negative reinforcement, that is realistic, tractable, and nontrapping?*

3. Perturbations of the Models

In this section, we add two features, noise and discounting, commonly used to create more realistic models. We examine the effects on social structure. In particular, these lead to varying degrees of subgroup formation.

Discounting the past

In the foregoing models, a positive (or negative) payoff in the distant past contributes equally to the weight (or resistance) assigned to an edge, as does a like payoff in the immediate past. This is implausible, both psychologically and methodologically. As a matter of psychology, memories fade. From the standpoint of inductive logic, it is not at all certain that the learner is dealing with stationary probabilities—indeed, in cases of prime interest, she is not. For this reason, recent experience may have a better chance of being a relevant guide to future action than the remote past.

A simple and standard way to modify the models to reflect this concern is to introduce discounting of the past. We will concentrate here on the models of Making Friends. After each interaction, we will now multiply the weights of the previous stage by a discount factor, d, between 0 and 1. The we add the undiscounted payoffs from the present interaction to get new weights. The modification of the dynamics has a dramatic effect on the Making Friends models.

For Friends I, it is immediately evident from simulations with $d = 0.9$, say, and ten players, that the probabilities p_{ij} converge to 0 or 1. In other words, each individual ends up always visiting the same other individual.

In Friends II, simulations show the group breaking into pairs, with each member of a pair always visiting his or her "partner." Which pairs form depends on the randomness in the early rounds of visits, but pairs always form. In fact, there are other possible limit states, but their frequency is low except at more extreme discount rates. The set of possible limit states may be described as follows. Some agents are grouped in pairs, each member of a pair always visiting the other. Other agents are grouped in *stars*. These are clusters of size at least three, in which one agent, called the *center*, visits each of the others with positive frequency, while the others always visit the center.

Analysis of discounting the past

It is worth giving a rigorous derivation of the above behavior, because it will shed some light on a defect in the most obvious log-likelihood model to incorporate positive and negative reinforcement. Our derivation highlights this, although the results for discounted Friends I may also be derived from a theorem of H. Rubin (see Davis 1990: 227).

THEOREM 4. *In Friends II with discount rate $d < 1$, there is always a partition into pairs and stars and a random time after which each member of a pair visits only the other member of the pair and each noncentral member of a star visits only the center. In Friends I, there is a random function f and a random time after which each player i always visits $f(i)$.*

Sketch of Proof: The analysis for Friends I is similar but easier, so we prove the statement only for Friends II. With each probability matrix p we associate a graph $G(p)$ as follows. The edge (i, j) is in the graph G if the probability $p_{ij} > \epsilon$, where $\epsilon < 1/(2n)$ is some fixed positive number. Among those graphs having at least one edge incident to each vertex, let S denote the minimal such graphs, that is, ones for which deleting any edge results in an isolated vertex. It is easy to see that S is the set $G(p)$ for all p satisfying the conclusion of the theorem.

The principle behind the analysis of discounted Friends is that the future behavior of p is largely determined by the present $G(p)$. In particular, we find a $\delta > 0$ such that from any state p, for each subgraph H of $G(p)$ such that $H \in S$, there is a probability at least δ^2 that for all sufficiently large t, $G(p(t)) = H$. We show this in two steps: (step *i*) with probability at least δ, there is some t for which $G(p(t)) = H$; (step *ii*) from any state p such that $G(p) = H$, there is probability at least δ that $G(p(t))$ is equal to H for all later times, t.

To see why step *i* is true, for $H \in S$, let f_H be any function on vertices of H for which each value $f(i)$ is a neighbor of i. Observe that there is a number k such that from any state p with $H \subseteq G(p)$, if each vertex i visits $f(i)$ for the next k rounds, then $G(p(k)) = H$. For each round of visits, this probability is at least ϵ^n, where n is the number of vertices, so taking $\delta \leq \epsilon^{kn}$ establishes (step *i*). For (step *ii*), it suffices to show that with probability δ each agent visits a neighbor in H at all later times. For each agent i, the sum over j not neighboring i in H of p_{ij} is at most $n\epsilon < 1/2$ by the definition of $G(p) = H$. After k rounds of visits where agents only visit their neighbors in H, this must decrease to at most $(1/2)d^k$. Thus the probability of N rounds of visits only to neighbors in H is at least

$$\prod_{k=0}^{N-1} \left(1 - \frac{1}{2}d^k\right)^n.$$

Sending N to infinity yields a convergent infinite product, since $(1/2)d^k$ is summable. Taking δ to be less than the infinite product proves (step *ii*).

With steps i and ii, the rest is a standard tail argument. The constraints on evolution are such that $G(p(t))$ always contains at least one graph in S. As long as it contains more than one graph in S, there is always a probability of at least δ of permanently settling into each one. Thus, with probability 1, eventually $G(p(t))$ is equal to some $H \in S$ for all future times. This is equivalent to the conclusion of the theorem.QED

Remark: It is actually shown that in (step ii), if we choose ϵ sufficiently small, we can choose δ arbitrarily close to 1.

We now also see why the log-likelihood rule (3) leads to fixation of a degenerate structure. Under these dynamics, an equivalent phenomenon occurs to step i in the proof of *Theorem 4*. For a pair (i, j) whose interaction has a positive mean, if the pair plays repeatedly, we will see $w_{ij}(t)/t \rightarrow \mu > 0$. The probability that i will ever switch partners, once having tried j a few times is at most on the order of $\sum_{k=0}^{\infty} B \exp(-k\mu)$, where $B = \exp\left(\sum_{l \neq j} w_{il}\right)$. From here it is easy to construct an argument parallel to the proof of *Theorem 4*, to show that in the presence of a game with positive mean payoff, discounted structural dynamics lead with probability 1 to fixation at a pairing.

Introduction of noise

A common feature in models of adaptation is the introduction of noise: a small chance of a behavior other than the one chosen by the dynamical equation for the model. This may stem from an agent's uncertainty, from agent error, or from circumstances beyond an agent's control. Alternatively, an agent may purposefully add noise to her strategy in order to avoid becoming wedded to a less than optimally efficient strategy or structure.

From a methodological point of view, noise that does not go to zero with time transforms the model into an ergodic Markov chain. No state is then trapping. To the extent that the trapping states produced by discounting or linear log-likelihood are unrealistic, we may hope to mitigate the problem by adding a noise component. Because dynamics with a noise term do not lead to a single state, the outcome is usually phrased in terms of *stochastically stable states* (Foster and Young 1990). A state is termed stochastically stable if the chance of finding the system near that state does not go to zero as the magnitude of the noise term goes to zero.

Neither discounting nor noise will affect the limiting behavior of Making Enemies. For Making Friends, let us modify the probability rule given in Eq. (1.1) so that in the n-player game, the probability of *i* visiting *j* is now some fixed positive number $\epsilon/(n-1)$, plus $(1-\epsilon)$ times what it was before:

$$p_{ij} = \frac{\epsilon}{n-1} + (1-\epsilon)\frac{w_{ij}}{\sum_k w_{ik}}.$$

The effect of this is to push the system by ϵ toward the uniform point $\bar{p}$. Neither Friends I nor Friends II is now a martingale, and the stable set of each is reduced to the single point $\bar{p}$. Because this is true at any noise level $\epsilon > 0$, we see that there is only one asymptotically stable point. Because the qualitative outcome is sensitive to the existence of a noise term, it is incumbent to ask with regard to specific models whether a noise term is natural and realistic.

Noise and discounting

In the presence of a discount $d < 1$ and a noise term $\epsilon > 0$, if $1-d$ is much smaller than ϵ, then the discount is so low that the noise term wipes out any effect the discounting might have had. In the other case, where d is held fixed and ϵ tends to zero, we may ask about the asymptotically stable states of a system with past discounting dynamics. For Friends I, nothing much interesting happens: discounting causes the limiting state to be degenerate; with noise, the system may jump from one such state to the other, which does not change which states are stochastically stable.

For Friends II, as long as the number of players *n* is at least 4, the introduction of noise does indeed change the set of stochastically stable states: it gets rid of stars. Simulations show that pairings are by far the most prevalent states in discounted Friends II, with a star of size 3 forming when necessitated by an odd number of players. We now show that states with more than one star, or a star of size greater than 3, are not stochastically stable.

THEOREM 5. *In Friends II, with discounting, with n players, and with noise tending to zero, the stochastically stable states are those that are either unions of pairs (if n is even) or pairs plus a single star of size 3 (if n is odd).*

Sketch of Proof: Let *S* denote the graphs corresponding to possible limit states as in the proof of *Theorem 4*, and let $S_0 \subseteq S$ denote those graphs with no stars (perfect pairings) or with a single star of size 3. The important properties of the relation of *S* to S_0 are as follows. (Property *i*) If *G* is the result of adding a single edge to a graph in S_0, then *G* contains no graph in $S \setminus S_0$. (Property *ii*) For any $G \in S$ there is a chain $G = G_1, G_2, \ldots, G_k$ leading to S_0, where each G_{j+1} may be obtained from G_j by adding an edge and then deleting two edges. Property i is apparent. To verify property ii, note that if $H \in S$ and i and j are noncentral vertices in stars of H, and they are not both in the same star of size 3, then adding the edge between i and j and removing the two edges previously incident to i and j produces a new graph in S. Iterating this procedure starting from $H = G_1$ leads in finite time (because the number of edges decreases each time) to an element of S_0.

We now follow the usual method for determining stochastic stability (33). Let the probability ρ of disobeying the structural dynamics Eq. (1.1) be very small. If ϵ (in the definition of S) is very small, then a state p with $G(p) = G \in S$ will have $G(p(t)) = G$ for all later times with high probability, until there is a disobeying move. After a single disobedience, the graph $G(p)$ will be the union of G with one extra edge. By the remark after the proof of *Theorem 4*, we see that after a disobedience, the graph will then relax to some subgraph in S. By property i, if $G \in S_0$ then this subgraph is again in S_0. Thus, a single disobedience followed by relaxation back to S will never escape S_0. Hence, the probability of jumping to $S \setminus S_0$ is of order ρ^2, which implies that states in S_0 stay in S_0 for time at least ρ^{-2}. On the other hand, by property ii, from any state in $S \setminus S_0$, there is a chain of single disobediences, such that allowing the system to relax after each may with positive probability land you back in S_0. Thus, the expected time spent in $S \setminus S_0$ before returning to S_0 is at most of order ρ. Thus, the process spends $(1 - \rho)$ portion of the time in S_0, and sending ρ to zero, we see that only states in S_0 are stochastically stable. It is easy to see that all of these are indeed stochastically stable.

4. Reinforcement by Games of Nontrivial Strategy

So far we have only considered a baseline model of uniform reinforcement, which turned out still to have nontrivial structural behavior. Now we examine a reinforcement scheme resulting from the payoff of a nontrivial game. We will consider the case where evolution of strategy is slower than evolution of structure. Thus, we will consider the agents as divided into types, each type always playing a fixed strategy, and see what sort of interaction structure emerges. We then extend this by allowing strategic switching of types. We find that coordination of strategy occurs, though whether players coordinate on the risk-dominant or payoff-dominant strategy depends on parameters of the model such as the rate of strategic evolution. Depending on conditions of the model, the social network may or may not split up into pairs.

Rousseau's Stag Hunt

Consider a two-player version of Rousseau's (1984) *Stag Hunt*. The choices are either to hunt stag or to hunt rabbit (hare, in the original). It takes two persons cooperating to effectively hunt a stag, while one person acting independently can hunt a rabbit. Bagging a stag brings a greater payoff.

	Hunt Stag	Hunt Rabbit
Hunt Stag	(1, 1)	(0, .75)
Hunt Rabbit	(.75, 0)	(.75, .75)

There are two equilibria in this game: both hunt stag and both hunt rabbit. The first carries the higher payoff and is said to be *payoff dominant*; the second carries the least risk and is said to be *risk dominant* (Harsanyi and Selten 1988). In models without structural dynamics, Kandori, Mailath and Rob (1993) have shown that only the risk-dominant equilibrium of a two-player coordination game is stochastically stable. In the presence of structural dynamics, we will describe a more optimistic conclusion.

THEOREM 6. *Suppose Stag Hunt is played by 2n players, with structural dynamics given by Eq. (1.1) and cumulative weighting dynamics (Eq. (1.2)) with no noise or discounting. Then in the limit, stag hunters always visit stag hunters and rabbit hunters visit rabbit hunters.*

Sketch of Proof: First note that no visit of a stag hunter to a rabbit hunter is ever reinforced. Thus, $w_{ij}(t) = 1$ for all t if i is a stag hunter and j is a rabbit hunter. Observing that the weights $w_{ij}(t)$ go to infinity when i and j are both stag hunters, we see that the probability of a stag hunter visiting a rabbit hunter goes to zero.

Next, consider the subpopulation of rabbit hunters, call it A. For $i \in A$, let

$$Z(i, t) = \frac{\sum_{j \notin A} w_{ij}}{\sum_{j=1}^{n} w_{ij}}$$

denote the probability of visiting a given rabbit hunter visiting a stag hunter on the next turn. The expected value of $Z(i, t + 1)$ changes according to the formula

$$E(Z(i, t + 1)|Z(i, t)) = Z(i, t) + t^{-1} Y(i, t),$$

where $Y(i, t)$ is the proportion of increase in expected weight w_{ij} due to $j \notin A$:

$$Y(i,t) = \frac{\sum_{j \notin A} p_{ij} + p_{ji}}{\sum_{j=1}^{n} p_{ij} + p_{ji}}.$$

Ignoring the terms p_{ji} in both the numerator and denominator of the above expression would lead to exactly $Z(i, t)$. The terms p_{ji} for $j \notin A$ are known to be small, while the total from the terms p_{ji} for $j \in A$ cannot be small. Consequently, $Y(i, t) < (1 - \epsilon)Z(i, t)$ for some $\epsilon > 0$, whence

$$E(Z(i, t+1) - Z(i, t) \mid Z(i, t)) \leq -\frac{\epsilon Z(i, t)}{t}.$$

Because the increments in $Z(i, t)$ are bounded by C/t, there are a $\lambda, \mu > 0$ for which $\exp(\lambda Z(i, t) + \mu \log t)$ is a supermartingale, which implies that $Z(i, t)$ converges to zero exponentially fast in $\log t$.QED

Introduction of a discount rate changes this outcome. Stag hunters still end up visiting stag hunters, because even discounted reinforcement beats a reinforcement of zero, but now rabbit hunters will get locked either into pairs and stars as in Making Friends, or into repeated visits to a single stag hunter. These limit states are all invariant under introduction of noise. When a rabbit hunter visits a stag hunter the loss to society is the 0.75 that another rabbit hunter would have profited from the visit. The model is evidently weak here, because it allows only one visit *by* each agent but any number of visits *to* each agent in a round of visits. That is, a more realistic loss would be the stag hunter's wasted time when visited by the rabbit hunter.

It should be noted that, although the stochastically stable states include those that are not optimally efficient, the optimally efficient states (those states where rabbit hunters visit rabbit hunters) will have an edge. Because of the possibility of reciprocal reinforcement, it will be easier for a rabbit hunter to switch from visiting a stag hunter to visiting a rabbit hunter than *vice versa*. Second, when the discount rate is near 1, the model behaves like the undiscounted model for a long enough time that it is very unlikely for a rabbit hunter to get locked into visiting a stag hunter in the first place. Simulations of Stag Hunting with ten players and $d = 0.9$, seem to show that rabbit hunters "always" visit rabbit hunters. Because of both of the effects mentioned above, the system is nearly always found in an optimally efficient state, even though there are stochastically stable states that are not optimally efficient.

Co-evolution of structure and strategy

To the previous model, we now add the possibility of an agent switching states: a stag hunter may decide to become a rabbit hunter, or a rabbit hunter may become bold and hunt stag. When this kind of strategic evolution is faster than the structural evolution, we know from studies of random encounter models that the risk-dominant equilibrium of everyone hunting rabbits will be arrived at while the network is still near its initial state of uniform visitation probabilities.

Whether strategic dynamics are faster or slower than structural dynamics depends, of course, on the activity being modeled; sometimes interaction structure is externally imposed, while sometimes it is more easily modified than strategy or character. Let us suppose that the investment in re-training as a different kind of hunter is great, so between each round of visits there is only a small chance that one of the hunters will change types. Then we have seen that hunters always (with no noise or discounting) or nearly always (in discounted models) hunt with others of like type. This eliminates the risk inherent in random encounters and allows hunters to profit from switching to stag hunting after an initial period where they find another stag hunter. Slow strategic adaptation gradually converts rabbit hunters to stag hunters and the payoff-dominant strategy dominates.

We describe here the results of simulations of Stag Hunting for 1,000 time steps, where with some probability q at any given time, an individual changes type to whichever type was most successful in the previous round. When $q = 0.1$, we found that in 22 percent of the cases all hunters ended up hunting stag, while in 78 percent of the cases, all hunters hunted rabbit. Thus there was perfect coordination, but usually not to the most efficient equilibrium. On the other hand, when $q = 0.01$, the majority (71 percent) of the cases ended in the optimal state of all hunting stag, while 29 percent ended up all hunting rabbit. Increasing the initial edge weights made it far less likely to reach the stag hunting equilibrium, since stag hunters took a long time to perfectly align, and without alignment, the previous round's best strategy was almost always rabbit hunting. For instance, if the initial weights were 1,000 for each visit, under 1 percent of the cases ended up all stag hunting, whether q was 0.1 or 0.01.

Once hunters largely cease to visit hunters of opposite type, the structural evolution within each of the two subpopulations is a version of Friends II. The resulting social structure will not be a perfect pairing, but will have each rabbit (stag) hunter visiting each other rabbit (stag) hunter, but with varying probabilities.

5. Conclusion

We have taken some basic steps in exploring the dynamics of the evolution of interaction structures and co-evolution of structure and strategy. The ultimate goals are to create models that are more true to life, and to find theoretical bases for observed behaviors of systems, including prediction of selection between multiple equilibria.

The particular dynamics we use here are only examples, but it turns out that the simplest of these may deliver interesting and surprising results. Even in baseline models where the game being played is degenerate, we find spontaneous emergence of structure from uniformity and spontaneous emergence of uniformity from structure. We find processes with extremely long transient modes, where limiting behavior is not a good guide for predicting behavior after thousands of trials.

The social interaction structures that emerge tend to separate the population into small interaction groups within which there is coordination of strategy. This separation may be complete, as in discounted Friends II, or may be only a tendency, as in the nondiscounted versions of Friends and Stag Hunting.

When we combine structure and strategy dynamics for a non-trivial game, the Stag Hunt, we find that the probable outcomes depend on the timing. Where structure is frozen in a random encounter configuration we get the expected risk-dominant equilibrium outcome. But when structure is fluid relative to strategy, structural adaptation neutralizes the risk and we get the socially efficient payoff-dominant equilibrium. Varying between these extremes can give one or the other result with different probabilities, or may leave the group in a state where both strategies are used. We expect to see structure dynamics making a difference in other games as well. Indeed, we have some preliminary simulation evidence showing this to be true for a bargaining game ("split the dollar"), and for a simple coordination game.

There are many more avenues to pursue. As mentioned in Making Enemies, it would be desirable to find a model in which positive and negative reinforcement are present, but trapping does not occur. We have not modeled any interaction among three or more players. We also have yet to model any explicit interaction between strategy and structure: the choice of a partner to play with and a strategy to play against that partner need not be independent.

One could continue adding complexity so as to allow information to affect structural evolution, to include communication between players, and so forth. Our main point is this. Structural change is a common feature of the real world. A theory of strategic interaction must take account of it. There is a mathematically rich theory which develops relevant tools. We believe that explicit modeling of structural dynamics, and the interaction of structure and strategy, will generate new insights for the theory of adaptive behavior.

Acknowledgments

We wish to thank Persi Diaconis, Joel Sobel, and Glenn Ellison for bringing us together, for helpful discussions, and for greatly improving our awareness of the relevant literature. This work was supported in part by National Science Foundation Grant DMS 9803249.

References

Anderlini, L. and A. Ianni (1997) "Learning on a Torus." In *The Dynamics of Norms*, ed. C. Bicchieri, R. Jeffrey, and B. Skyrms, 87–107. Cambridge: Cambridge University Press.

Arthur, W. B. (1989) "Competing Technologies, Increasing Returns, and Lock-in by Historical Events." *Economic Journal* 99: 116–31.

Benaïm, M. and M. Hirsch (1995) "Dynamics of Morse–Smale urn processes." *Ergodic Theory Dynamic Systems* 15: 1005–30.

Blume, L. (1993) "The Statistical Mechanics of Strategic Interaction." *Games and Economic Behavior* 5: 387–423.

Davis, B. (1990) "Reinforced random walk." *Probability Theory and Related Fields* 84: 203–29.

Diaconis, P. and D. Stroock (1991) "Geometric Bounds for Eigenvalues of Markov Chains." *Annals of Applied Probability* 1: 39–61.

Durrett, R. and S. Levin (1994) "The Importance of Being Discrete (and Spatial)." *Theoretical Population Biology* 46: 363–94.

Durrett, R. and C. Neuhauser (1997) "Coexistence Results for some Competition Models." *Annals of Applied Probability* 7: 10–45.

Durrett, R. (1996) *Probability: Theory and Examples*, 2nd edn. Belmont, CA: Duxbury Press, Wadsworth Publishing Company.

Eggenberger, F. and G. Pólya (1923) "Uber die Statistik verketter vorgäge." *Zeit. Angew. Math. Mech.* 3: 279–289.

Ehrenfest, P. and T. Ehrenfest (1907) "Ueber zwei bekannte Einwande gegen das Boltzmannsche H -Theorem." *Physiche Zeitschrift* 8: 311–14.

Ellison, G. (1993) "Learning, Local Interaction, and Coordination." *Econometrica* 61: 1047–71.

Ellison, G. (2000) "Basins of Attraction, Long-Run Stochastic Stability, and the Speed of Step-by-Step Evolution." *Review of Economic Studies*.

Epstein, J. and R. Axtell (1996) *Growing Artificial Societies*. Cambridge, MA: MIT/Brookings.

Feldman, M. and E. Thomas (1987) "Behavior-Dependent Contexts for Repeated Plays in the Prisoner's Dilemma II: Dynamical Aspects of the Evolution of Cooperation." *Journal of Theoretical Biology* 128: 297–315.

Foster, D. and H. P. Young (1990) "Stochastic evolutionary game dynamics." *Theoretical Population Biology* 38: 219–32.

Harsanyi, J. and R. Selten (1988) *A General Theory of Equilibrium in Games*. Cambridge, MA: MIT Press.

Hofbauer, J. and K. Sigmund (1988) *The Theory of Evolution and Dynamical Systems*. Cambridge: Cambridge University Press.

Jackson, M. and A. Watts (1999) "On the Formation of Interaction Networks in Social Coordination Games." *Games and Economic Behavior* 41: 265–91.

Johnson, N. and S. Kotz (1977) *Urn Models and Their Application*. New York: Wiley.

Kac, M. (1947) "Random Walk and the Theory of Brownian Motion." *American Mathematical Monthly* 54: 369–91.

Kandori, M., G. Mailath, and R. Rob (1993) "Learning, mutation, and long-run equilibria in games." *Econometrica* 61: 29–56.

Kang, H.-C., S. Krone, and C. Neuhauser (1995) "Stepping-stone Models with Extinction and Recolonization." *Annals of Applied Probability* 5: 1025–60.

Lindgren, K. and M. Nordahl (1994) "Evolutionary Dynamics of Spatial Games." *Physica D* 75: 292–309.

Pemantle, R. (1990) "Nonconvergence to unstable points in urn models and stochastic approximations." *Annals of Probability* 18: 698–712.

Pollack, G. B. (1989) "Evolutionary Stability in a Viscous Lattice." *Social Networks* 11: 175–212.

Robbins, H. and S. Monro (1951) "A stochastic approximation method." *Annals of Mathematical Statistics* 22: 400–7.

Rousseau, J.-J. (1984) *A Discourse on Inequality*; trans. Cranston, M. London: Penguin.

Schelling, T. (1969) "Models of Segregation." *American Economic Review: Papers and Proceedings* 59: 488–93.

Schelling, T. (1971) "Dynamic Models of Segregation." *Journal of Mathematical Sociology* 1: 143–86.

Talagrand, M. (1995) "Concentration of Measure and Isoperimetric Inequalities in Product Spaces." *IHES Publications Mathématiques* 81: 73–205.

Watts, D. and S. Strogatz (1998) "Collective dynamics of 'small-world' networks." *Nature (London)* 393: 440–2.

Weibull, J. (1997) *Evolutionary Game Theory*. Cambridge, MA: MIT Press.

Wright, S. (1921) "Systems of Mating III: Assortative Mating Based on Somatic Resemblance." *Genetics* 6: 144–61.

Wright, S. (1945) "Tempo and Mode in Evolution: A Critical Review." *Ecology* 26: 415–19.

Zajonc, R. B. (1968) "Attitudinal Effects of Mere Exosure." *Journal of Personality and Social Psychology Monograph* 9, 1–28.

12

Network Formation by Reinforcement Learning: The Long and Medium Run

with Robin Pemantle

1. Introduction

Each day, each member of a small group of individuals selects two others with whom to interact. The individuals are of various types, and their types determine the payoff to each from the interaction. That is to say that the interaction is modeled as a symmetric three-person game. Probabilities of selecting individuals evolve by reinforcement learning, where the reinforcements are the payoff of the interaction. We consider two games. The first is a degenerate game, "Three's Company". Here there is only one type and everyone gets equal reinforcement for every interaction. The analysis of "Three's Company" is then used in the analysis of a second game, a three-person Stag Hunt. Here there are two types, Stag Hunters and Hare Hunters. Hare Hunters always get a payoff of 3, but a Stag Hunter gets a payoff of 4 if he interacts with two other Stag Hunters, otherwise he gets nothing.

We do not pretend to be giving a realistic account of the friendship formation or of the small group dynamics of hunting. But we hope that our modeling exercise has a significance that is more than purely philosophical. This hope is based on three considerations. First, we use a model of reinforcement learning that is backed by a large body of laboratory data for both animals and humans, including interactive human learning in games. In analyzing the models we focus on ranges of parameter values of the learning dynamics that have been found in the experimental literature. Second, the notion of modeling the co-evolution of interaction networks and strategies was set out in Skyrms and Pemantle (2000). This calls for the formulation and analysis of the most basic stochastic network models. The idea is that basic, simple models can serve as modules for the construction of more complex models. To this end, we provide an analysis of two simple models. As the second model evolves, the first model appears as a module. Stag Hunters end up playing "Three's Company" among themselves. Finally, we aim at a robust analysis. Our analysis of "Three's Company" and "Stag Hunt" is sufficiently robust to shed light on similar models, which share a common mathematical description. We aim to provide a collection of rigorous results on this class of stochastic models, which will help scientists to understand the similarities and differences between their long- and medium-term behavior.

2. Reinforcement Learning and Reinforcement Processes

Reinforcement learning, as the term is used in psychology, and reinforcement models, as used in applied mathematics, are not coextensive. We will model reinforcement learning by a certain kind of reinforcement process, following Herrnstein (1970) and Roth and Erev (1995). In a reinforcement process, as defined in the mathematical literature on reinforced random walks (Coppersmith and Diaconis 1997; Davis 1990, 1999; Pemantle 1988, 1992) the current probability can depend on the entire history of choices and payoffs, via summary statistics or propensities associated to the possible actions. The possibility of using such processes to model reinforcement learning was introduced by Luce (1959).

Luce considered a range of models for the evolution of the propensities. The payoffs for an action taken might modify its propensity multiplicatively, additively, or in some combination. He computes the probabilities of actions by simply normalizing the propensities. This is Luce's linear response rule. The separation of the questions of propensity evolution and response rule opens the possibility of other alternatives such as the logistic response rule used by Busemeyer and Townsend (1993) and by Camerer and Ho (1999).

Herrnstein (1970) quantified Thorndike's "Law of Effect," as the "Matching Law": the probability of choosing an action is proportional to the accumulated rewards. Let propensities evolve by adding the payoff of the action chosen to its propensity. If we follow Luce's linear response rule, we obtain Herrnstein's matching law. Herrnstein reports data from laboratory experiments with humans as well as with animals, which support the broad applicability of the model.

There is a special case whose limiting behavior is well-known. If each action is equally reinforced, the process is mathematically equivalent to Pólya's urn process (Eggenberger and Pólya 1923), with each action represented by a different color of ball initially in the urn. The process converges to a random limit, whose support is the whole probability simplex. In other words, any limiting state of propensities or probabilities is possible.

In a pioneering study, Suppes and Atkinson (1960) used a different (Markovian) model of reinforcement learning to model learning behavior in games. A number of players choose between alternatives as before, but the payoffs to each player now depend on the acts chosen by all players. Players modify their choice probabilities by the learning dynamics.

In 1995, Roth and Erev (1995) proposed a multi-agent reinforcement model based on Herrnstein's linear reinforcement and response. Here and in subsequent publications, Erev and Roth (1998) and Bereby-Meyer and Erev (1998) show a good fit with a wide range of empirical data. Limiting behavior in the basic model has recently been studied by Beggs (2002).

In Skyrms and Pemantle (2000), both basic and discounted versions of Roth–Erev learning are applied to social network formation. Individuals begin with prior propensities to interact with each other, and interactions are modeled as two-person games. Individuals have given strategies, and interactions between individuals evolve by reinforcement learning. The analysis begins with a series of results on "Making Friends," a network formation model in the special case where the game interaction is trivial. Nontrivial strategic interaction is then introduced, and it is shown that the co-evolution of network and strategy depends on relative rates of evolution as well as on other features of the model.

The present work is a natural outgrowth of the investigations begun in Skyrms and Pemantle (2000). In the richer context of multi-agent interactions, more phenomena arise, namely clique formation and a meta-stable state of high network connectivity for an initial epoch whose length depends dramatically on the discounting parameter. In Section 5.3 we discuss the implications of these features for a wide class of models.

3. Mathematical background

Our ultimate goal is to understand qualitative phenomena such as clique formation, or the tendency of the interaction frequencies toward some limiting values. The mathematical literature on reinforcement processes contains results in these directions. It will be instructive to review these, and to examine the mathematical classification of such processes, although we will need to go beyond this level of analysis to explain the behavior of network models such as Three's Company on timescales we can observe.

Reinforcement processes fall into two main types, trapping and non-trapping. A process is said to be trapping if there are proper subsets of actions for each player such that there is a positive probability that all players always play from this subset of actions. For example, if the repetition of any single vector (i) of actions (action i_j for player j) is sufficiently self-reinforcing that it might cause action i to be perpetuated forever, then the process is trapping. The specific dynamics investigated by Bush and Mosteller (1955) are trapping, as are most logistic response models. By contrast, models that give all times in the past an equal effect on the present, such as Herrnstein's dynamics and Roth–Erev dynamics, tend not to be trapping.

One of several modifications suggested by Roth and Erev to maximize agreement of their model with the data is to introduce a discounting parameter $x \in (0, 1)$. The past is discounted via multiplication by a factor of $(1 - x)$ at each step. It is known from the theory of urn processes that discounting may cause trapping. For example, it follows from a theorem of H. Rubin, reported in Davis (1990), that if Pólya's urn is altered by discounting the past, there will be a point in time beyond which only one color is ever chosen. This holds as well with Roth–Erev type models: the discounted Roth–Erev model is trapping, while the undiscounted model is not. In Skyrms and Pemantle (2000), discounted and nondiscounted versions of several games are studied, and equilibria examined for stability. Again, discounting causes trapping, and we investigate the robustness of the trapping when the discounting parameter is close to negligible. In a related paper, Bonacich and Liggett (2003) investigate Bush–Mosteller dynamics in a two-person interaction representing gift giving. Their model has discounting, and they find a set of trapping states.

It is in general an outstanding problem in the theoretical study of reinforcement models to show that trapping must occur with probability 1 if it occurs with positive probability. This was only recently proved, for instance, for the reinforced random walk on a graph with three vertices, via a complicated argument in Limic (2001). Much of the effort that has gone into the mathematical study of these models has been directed at these difficult limiting questions. In the non-trapping case, even though the choice of action does not fixate, the probabilities for some of the actions may tend to zero. A series of papers in the 1990s by Benaïm and Hirsch (1995; Benaïm 1998, 1999) establishes some basic tests for whether in undiscounted Roth–Erev type models, probabilities will tend toward deterministic vectors.

From the point of view of applications, in a situation where it can be proven or surmised that trapping occurs, we are mainly interested in characterizing the states in which we may become trapped and in determining how long it will be before the process becomes trapped. Recalling our initial discussion of modeling goals, we are particularly interested in results that are robust as parameters and modeling details vary or, when they are not robust, of understanding how these details of the model affect observed qualitative behavior.

The present work is a natural outgrowth of the investigations begun in Skyrms and Pemantle (2000). In the richer context of multi-agent interactions, more phenomena arise, namely clique formation and a meta-stable state of high network connectivity for an initial epoch whose length depends dramatically on the discounting parameter. In Section 5.3 we discuss the implications of these features for a wide class of models.

4. Three's Company: a ternary interaction model

4.1 Specification of the model

The game "Three's Company" models collaboration of trios of agents from a fixed population. At each time step, each agent selects two others with whom to form a temporary collusion. The two selected always agree to participate. And at each time step of the process—you can think of it as a day—there is ample opportunity for each individual to initiate a trio. Thus, an agent may be involved in multiple collusions during a single time step: one that she initiates, and zero or more initiated by the other agents. Analogously to the basic game "Making Friends", introduced in Skyrms and Pemantle (2000), Three's Company has a constant reward structure: every collaboration results in an identical positive outcome, so every agent in every temporary collusion increases by an identical amount her propensity to choose each of the other two agents in the trio. The choice probabilities follow what could be called multilinear response. The probability of an agent choosing to form a trio with two other agents i and j is taken to be proportional to the product of her propensity for i with her propensity for j. In addition to providing a model for self-organization based on a simple matching law type of response mechanism, this model is meant to provide a basis for the analysis of games such as the three-person stag hunting game discussed in the next section. We now give a more formal mathematical definition of Three's Company.

Fix a positive integer $N \geq 4$, representing the size of the population. For $t \geq 0$ and $1 \leq i, j \leq N$, define random variables $W(i, j, t)$ and $U(i, t)$ inductively on a common probability space $(\Omega, F, \mathbb{P})$ as follows. The W variables are positive numbers, and the U variables are subsets of the population of cardinality 3. One may think of the U variables as random triangles in the complete graph with a vertex representing each agent. The variable $U(i, t)$ is equal to the trio formed by agent i at time t. The W variables represent propensities: $W(i, j, t)$ will be the propensity for player i to choose player j on the time step t. The initialization is $W(i, j, 0) = 1$ for all $i \neq j$, while $W(i, i, 0) = 0$. We write $W(e, t)$ for $W(i, j, t)$ when e is the edge (unordered set) $\{i, j\}$ (note that the evolution rules below imply that $W(i, j, t) = W(j, i, t)$ for all i, j, and t). The inductive step, for $t \geq 0$, defines probabilities (formally, conditional probabilities given the past) for the variables $U(i, t)$ in terms of the variables $W(r, s, t)$, $r, s \leq N$, and then defines $W(i, j, t+1)$ in terms of $W(i, j, t)$ and the variables $U(r, t), r \leq N$. The equations are:

$$\mathbb{P}(U(i, t) = S \mid \mathcal{F}_t) = \frac{1_{i \in S} \Pi_{r,s \in S, r<s} W(r, s, t)}{\sum_{S': i \in S'} \Pi_{r,s \in S', r<s} W(r, s, t)}; \qquad (4.1)$$

$$W(i, j, t+1) = (1 - x) W(i, j, t) + \sum_{r=1}^{N} 1_{i,j \in U(r,t)}. \qquad (4.2)$$

Here $(1 - x)$ is the factor per unit time by which the past is discounted, and the σ-field conditioned on is the process up to time t,

$$\mathcal{F}_t := \sigma\{W(i, j, u) : u \leq t\}.$$

The following alternative statement of the evolution Eq. (4.1) is useful for those familiar with the analytic machinery, cf. Pemantle (1992), that is typically used to reduce such a process to a stochastic approximation. Think of the normalized matrix

$$W_t := \frac{1}{\sum_{i,j} W(i, j, t)} W(\cdot, \cdot, t)$$

as the state vector. This is then an asymptotically time-homogeneous Markov chain, with an evolution rule

$$\mathbb{E}(W_{t+1} - W_t \mid \mathcal{F}_t) = g(t)[\mu(W_t) + \xi_t] \qquad (4.3)$$

where $g(t) = x + O(1/t)$, the drift vector field μ maps the simplex of normalized matrices into its tangent space and may be explicitly computed, and ξ_t are martingale increments of order 1. In the non-discounted case, $g(t) = O(1/t)$, and much information about the long-term behavior of this process can be discovered by an analysis of the the flow $dX/dt = \mu(X)$ Benaïm (1999). In the discounted case, $g(t)$ does not go to zero and an alternate analysis is required.

4.2 Analysis of the model

Eqs. (4.1) and (4.2) completely specify the model for the given parameters N and x. Simulations for a population of size 6 ($N = 6$) showed the following behavior.

When $x = 0.5$ (a rather steep discount rate, though not unheard of in psychological laboratory experiments (Busemeyer and Stout 2002)), all 1000 trials broke up into two cliques of size 3, with no interactions across clique boundaries. In larger populations, with the same discount rate, again decomposition into cliques occurs, this time of sizes 3, 4, and 5, whose members interact exclusively with other members of the same clique.

When $N = 6$ and $x = 0.4$ we found that 994 out of the 1000 trials had decomposed into two cliques of three (we allowed the process to continue for 1,000,000 time steps). When x was decreased to 0.3, only 13 of the 1,000 trials showed decomposition into cliques, while in the remainder of the trials all six members of the population remained well connected through the 1,000,000 time steps. Finally, when $x = 0.2$, a reasonable discount rate for individuals though still steeper than in most economic models, none out of 1,000 trials had broken into cliques. All six members of the population remained well connected after 1,000,000 time steps.

To put these results in perspective, we note that a number of laboratory experiments have been used to estimate the discount rate that best fits the aggregate data for Roth–Erev learning in games played in the laboratory. The best estimates of x in the data discussed in Bereby-Meyer and Erev (1998) put x less than 0.01. There seems, however, to be a great deal of individual variability in discount rates with some individuals having a value of x near 0.5, and for humans in other settings or for other biological systems the discount rate could be quite different.

To summarize the simulation data, high discount rates lead to trapping, with each agent restricting her choices to members of a clique of size 3 (or, in larger populations, size 4 or 5). Less steep discount rates lead to less trapping or no trapping at all. Interestingly, the simulation data are contradicted by the following theorem, proved in Appendix A.

THEOREM 4.1. *In Three's Company, with any population size $N \geq 6$ and any discount rate $x \in (0, 1)$, with probability 1 the population may be partitioned into subsets of sizes 3, 4, and 5, such that each member of each subset chooses each other with positive limiting frequency, and chooses members outside the subset only finitely often. Every partition into sets of sizes 3, 4, and 5 has positive probability of occurring.*

In other words, despite the simulation data, trapping always occurs. The set of traps is the set of all ways of decomposing into cliques of sizes

3, 4, and 5. The apparent contradiction between the simulation and the theorem is resolved by *Theorem 4.2*, whose proof is given in the companion paper Pemantle and Skyrms (2004). The theorem states that the time for the population to break into cliques increases exponentially in $1/x$ as the discount rate $1 - x$ increases to 1.

THEOREM 4.2. *For each $N \geq 6$ there is a $d > 0$ and numbers $c_N > 0$ such that in Three's Company with N players and discount rate $1 - x$, the probability is at least d that each player will play with each other player beyond time $exp(c_N x^{-1})$.*

5. The Three-player Stag Hunt

5.1 Specification of the model

We now replace the uniformly positive reward structure by a nontrivial game, which is a three-player version of Rousseau's Stag Hunt. For the purposes of our model, agents are divided into two types, hare hunters and stag hunters. That is, we model strategic choice as unchanging, at least on the time scale where network evolution is taking place. No matter which other two agents a hare hunter goes hunting with, the hare hunter comes back with a hare (hares can be caught by individuals). A stag hunter, on the other hand, comes home empty-handed unless in a trio of three stag hunters, in which case each comes home with one-third share of a stag. One-third of a stag is better than a whole hare, but evidently riskier because it will not materialize if any member of the hunting party decides to play it safe and focus attention on bagging a hare. In the three-player Stag Hunting game, as in Three's Company, at each time step each agent chooses two others with whom to form a collusion. The payoffs are as follows. Whenever a hare hunter is a member of a trio, his reward is 3. A stag hunter's reward is 4 if in a trio of three stag hunters and 0 otherwise. A formal model is as follows.

Let $N = 2n$ be an even integer representing the size of the population and let $x \in (0, 1)$ be the discount parameter. The variables $\{W(i, j, t), U(i, t): 1 \leq i, j \leq N; t \geq 0\}$ are defined again on $(\Omega, F, \mathbb{P})$ with the W variables taking positive values and representing propensities and the U variables taking values in the subsets of $\{1, \ldots, N\}$ of cardinality 3 and representing choices of trios. We initialize the W variables by $W(i, j, 0) = 1 - \delta_{ij}$, just as before, and we invoke a linear response mechanism (Eq. (4.1)) just as before. Now, instead of the trivial reward structure (Eq. (4.2)), the propensities evolve according to the hunting bounties

$$W(i,j,t+1) = (1-x)W(i,j,t) + 31_{i\leq n}\sum_{r=1}^{N} 1_{i,j\in U(r,t)} + 21_{i>n}$$

$$\sum_{q,r,s=n+1}^{N} 1_{i\in U(q,t)=\{q,r,s\}}. \tag{5.4}$$

The factor in front of the last sum is 2 because the sum counts the trio $\{q, r, s\}$, chosen by agent q, exactly twice, as (q, r, s) and as (q, s, r).

5.2 Analysis of the model

The propensities for stag hunters to choose hare hunters remain at their initial values, whence stag hunters choose other stag hunters with limiting probability 1. The stag hunters are never affected by the hare hunters' choices, so the stag hunters mimic Three's Company among themselves precisely except for the times, numbering only $O(\log t)$ uniformly in x by time t, when they choose hare hunters. We know therefore, that eventually they fall into cliques of size 3, 4, and 5, but that this will take a long time if the discount parameter is small.

Hare hunters may form cliques of size 3, 4, and 5 as well, but because they are rewarded for choosing stag hunters, they may also attach to stag hunters. The chosen stag hunters have cliques of their own and ignore the hare hunters, except during the times that they are purposelessly called to hunt with them. These attachments can be one hare hunter continually calling on a particular pair of stag hunters or two hare hunters continually calling on a single stag hunter. In either case the one or two hare hunters are isolated from all hunters other than their chosen stag hunters.

What matters here is not the details of the trapping state but the time scale on which the trap forms and the likelihood of a hare hunter ending up in a sub-optimal trap.[2] This likelihood decreases as the discount rate becomes small for the following reason. Hare hunters choosing to hunt

---

[2] In this model, since any number of collusions is permitted for an agent on each time step, the sub-optimality is manifested not through wasted time on the stag hunter's part. Instead, it is a societal opportunity cost, borne by all the hare hunters passed over in favor of the stag hunter.

with stag hunters are getting no reciprocal invitations, whereas any time they choose to hunt with other hare hunters, their mutual success creates a likelihood of future reciprocal invitations. These reciprocal invitations are then successful and increase the original hunter's propensity for choosing the other hare hunter. Thus, on average, propensity for a hare hunter to form a hunting party with other hare hunters will increase faster than propensity to call on stag hunters, and the relative weights will drift toward the hare–hare groupings. The smaller the discount parameter, x, the more chance this has to occur before a chance run of similar choices locks an agent into a particular clique.

Simulations show that stag hunters find each other rapidly. With six stag hunters and six hare hunters and a discount rate of 0.5, the probability that a stag hunter will visit a hare hunter usually drops below half a percent in 25 iterations. For 50 iterations of the process this always happened in 1000 trials, and this remains true for values of x between 0.5 and 0.1. For $x = 0.01$, 100 iterations of the process suffices for stag hunters to meet stag hunters at this level and for 200 iterations are enough when $x = 0.001$. Hare hunters find each other more slowly, except when they are frozen into interactions with stag hunters. When the past is heavily discounted the latter possibility is a serious one. At $x = 0.5$, at least one hare hunter interacted with a stag hunter (after 10,000 iterations) in 384 of 1,000 trials. This number dropped to 217 for $x = 0.4$, 74 for $x = 0.3$, 6 for $x = 0.2$, and 0 for $x = 0.1$. Reliable clique formation among stag hunters is much slower in line with the results of the last section, taking about 100,000 iterations for $x = 0.5$ and 1,000,000 iterations for $x = 0.6$.

5.3 General principles

The two models discussed in this chapter are highly idealized. But from these, we can learn some general principles as to how to analyze a much wider class of models.

When x is near zero, the process should for a long time behave similarly to the non-discounted process ($x = 0$). To understand non-discounted process, following Benaïm (1999), one must find equilibria for the flow $dX/dt = \mu(X)$, and classify these as to stability. Unstable equilibria, in general, do not matter. The nondiscounted process cannot converge to an unstable equilibrium of the flow. (However, for cases in which the effects of unstable equilibria may last quite a while, see Pemantle and Skyrms (2003).)

Stable equilibria of the flow may or may not be possible trapping states of the discounted stochastic process. Perhaps the most interesting case is that illustrated by Three's Company, where a stable equilibrium for the flow is not a possible trapping state for the discounted process. In Three's Company with six individuals, the state in which each individual chooses each possible pair of companions with equal probability is a stable equilibrium of the flow. As we have seen, it is not a trapping state of the discounted process. The discounted process may get pseudo-trapped there, that is, may remain there for a very long time.

This is a general phenomenon. When there are stable states of the flow that are non-trapping for the discounted process such psuedo-trapping may occur. Regarding this phenomenon, Theorem 4.2 extends rather robustly to a broader class of linearly stable states (for the flow corresponding to the non-discounted process) that are non-trapping for the discounted process (Pemantle and Skyrms 2004).

6 Conclusion

Our analysis reinforces the emphasis of Suppes and Atkinson, and of Roth and Erev, on the medium run for empirical applications. Long-run limiting behavior may simply never be seen. It is useful to quantify the time scale on which we can expect medium-run behavior to persist, and Theorem 4.2 is meant to serve as a prototypical result in this direction. Indeed, Theorem 4.2 is proved via a stronger result (Pemantle and Skyrms 2004: Theorem 4.1), which applies to many trapping models as the discount rate becomes negligible. As to the nature of the medium-run behavior, analyses tend to be model-dependent.

Appendix A. Proof of Theorem 4.1

Let $g(t)$ be the graph whose edges are all e such that $e \subseteq U(i, t)$ for some i, that is, the set of edges whose weights are increased from time t to time $t + 1$. The following two easy lemmas capture some helpful estimates.

LEMMA 7.1. *(i)*

$$\sum_e W(e,t) \to 3Nx^{-1}$$

exponentially fast as $t \to \infty$
(ii) If $e \in G(t)$ then

$$W(e, t+k) \geq (1-x)^{k-1}.$$

Proof. The first part is a consequence of the equation for the evolution of the total weight:

$$\sum_e W(i, t+1) = (1-x) \sum_e W(e,t) + 3N.$$

The second part follows from the first, and from the fact that when $e \in G(t)$ then $W(e, t+1) \geq 1$ and hence $W(e, t+k) \geq (1-x)^{k-1}$.

Let $\bar{G}$ denote the transitive (irreflexive) closure of a graph G; thus $\bar{G}$ is the smallest disjoint union of complete graphs that contains G. $\square$

LEMMA 7.2. *There are constant c, depending on N and x, such that with probability at least c, every edge $e \in \overline{G(t)}$ satisfies*

$$W(e, t+N^2) \geq (1-x)^{N^2}. \tag{7.1}$$

Proof. Let H be a connected component of $G(t)$. Fix a vertex $v \in H$ and let w be any other vertex of H. There is a path from v to w of length at most N; denote this path $(v = v_1, v_2, \ldots, v_r = w)$. If $r = 2$ then the inequality (Eq. (7.1)) for $e \in G(t)$ follows from Lemma 7.1. If $r \geq 3$, we let $E(H, v, w, 1)$ be the event that for every $2 \leq j \leq r-1$, the edge $\{v_{j-1}, v_{j+1}\}$ is in $G(t+1)$. Since this event contains the intersection over r of the events that $U(v_j, t) = \{v_j, v_{j-1}, v_{j+1}\}$, since Lemma 7.1 bounds each of these probabilities from below, and since the events are conditionally independent given F_t, we have a lower bound on the probability of $E(H, v, w, 1)$. In general, for $1 \leq k \leq r-2$, let $E(H, v, w, k)$ be the event that for every $2 \leq j \leq r-k$, the edge $\{v_{j-1}, v_{j+k}\}$ is in $G(t+k)$. We claim that conditional on $E(H, v, w, l)$ for all $l < k$, the conditional probability of $E(H, v, w, k)$ given F_{t+k-1} can be bounded below: inductively, Lemma 7.1 bounds from below the product of $W(v_j, v_{j-1}, t) W(v_j, v_j + k, t)$, and hence the probability that $U(v_j, t) = \{v_j, v_{j-1}, v_{j+k}\}$; these conditionally independent probabilities may then be multiplied to prove the claim, with the bound depending only on x and N.

From this argument, we see that the intersection $E(H, v, w) := \cap_{1 \leq k \leq r-2} E(H, v, w, k)$ has a probability which is bounded from below. Sequentially, we may choose a sequence of values for w running through all vertices of H at some distance $r(w) - 1 \geq 2$ from v, measured in the metric on H. For each such w, we can bound from below the probability that in $r - 2$ more time steps the path from v to w will be transitively completed. We denote these events $E'(H, v, w)$, the prime denoting the time shift to allow events analogous to $E(H, v, w)$ to occur sequentially. Summing the time to run over all $w \in H$ yields at most N^2 time steps. Let $E(H, v)$ denote the intersection of all the events $E'(H, v, w)$. Inductively, we see that the probability of $E(H, v)$ is bounded from below by a positive number depending only on N and x.

Finally, we let (H, v) vary with H exhausting components of $G(t)$ and v a choice function on the vertices of H. The events $E(H, v)$ are all conditionally independent given F_t, so the probability of their intersection, E, is bounded from below by a positive constant which we call c. By Lemma 7.1 once more, on E, we know that Eq. (7.1) is satisfied for each $e \in \overline{G(t)}$. □

Proof of Theorem 4.1. For any subset V of agents, let

$$E(V, t) := \bigcap_{s \geq t} \bigcap_{v \in V, w \in V^c} \{\{v, w\} \notin G(s)\}$$

denote the event that from time t onward, V is isolated from its complement. If V is the vertex set of a component of $G(t)$, then the conditional probability given F_t of the event $E(V, t)$ may be bounded from below as follows. For any $v \in V$, $w \in V^c$, and for any $s \geq t$, if the edge $e := \{v, w\}$ is not in $G(r)$ for any $t \leq r < s$, then by part 1 of Lemma 7.1, its weight $W(e, s)$ is at most $(1 - x)^{s-t} 3Nx^{-1}$. Since $\Sigma_z W(v, z, s) \geq 2$ for all v, z, s, it follows from the evolution equations that

$$\mathbb{P}(e \in G(s) \mid F_s) \leq \frac{(1 - x)^{s-t} 3Nx^{-1}}{2 + (1 - x)^{s-t} 3Nx^{-1}}.$$

It follows that

$$\mathbb{P}\left(\exists v \in V, w \in V^c : \{v, w\} \in G(s) + \mathcal{F}_s\right) = O(Nx^{-1}(1 - x)^{s-t})$$

uniformly in N, x and t as $s - t \to \infty$ (though the uniformity in N and x is not needed). By the conditional Borel–Cantelli Lemma, it follows that

$$\mathbb{P}(E(V, t) \mid F_t) > c(N, x) \tag{7.2}$$

on the event that V is the vertex set of a component of $\overline{G(t)}$.

By the reverse direction of the Conditional Borel–Cantelli Lemma, the event $E(V, t)$ occurs for some t with probability 1 on that event that V is a component of $\overline{G(t)}$ infinitely often. Let $e = \{v, w\}$ be any edge. If $e \notin \overline{G(t)}$ infinitely often, then since there are only finitely many subsets of vertices, it follows that $v \in V$ and $w \in W$ for some disjoint V and W that are infinitely often components of $G(t)$. This implies that $e \in \overline{G(t)}$ finitely often. We have shown that, almost surely, the edges come in two types: those in $\overline{G(t)}$ finitely often and those in $\overline{G(t)}$ all but finitely often. This further implies that $\overline{G(t)}$ is eventually constant. Denote this almost sure limit by G_∞. It remains to characterize G_∞.

It is evident that G_∞ contains no component of size less than three, since $G(t)$ is the union of triangles $U(i, t)$. Suppose that $\overline{G(t)} = H$ for some H of cardinality at least six. By Lemma 7.2, conditional on F_t and $\overline{G(t)} = H$,

$$W(e, t + N^2) \geq \frac{(1 - x)^{N^2}}{3N}$$

for every $e \in H$. Write H as the disjoint union of sets J and K, each of cardinality at least three. Then with probability at least

$$\left(\frac{(1 - x)^{N^2}}{3N + (1 - x)^{N^2}} \right)^{|J| + |K|}$$

$U(i, t + N^2) \subseteq J$ for every $i \in J$ and $U(i, t + N^2) \subseteq K$ for every $i \in K$. In this case, $\overline{G(t + N^2)} H$ has components that are proper subsets of H. By the martingale convergence theorem,

$$\mathbb{P}(H \text{ is a component of } G_\infty \mid \mathcal{F}_t)$$

converges with probability 1 to the indicator function of H being a component of G_∞. From the above computation, it is not possible for $\mathbb{P}(H \text{ is a component of } G_\infty | F_t)$ to converge to 1 when H has cardinality six or more. Therefore, every component of G_∞ has cardinality 3, 4, or 5.

The rest of the proof is easy. Let $V_1, \ldots, V_k$ be any partition of $[N]$ into sets of cardinalities 3, 4 and 5. The derivation of Eq. (7.2) shows that

$$P\left(\bigcap_{j=1}^{k} E(V_j, 1) \right) > 0,$$

in other words, with positive probability G_∞ has k components which are precisely the complete graphs on $V_1, \ldots, V_k$. It is elementary that a coupling may be produced between the Three's Company processes on populations of sizes N and $K < N$ (with the same x value), so that if $\{ \tilde{W}(i, j, t), \tilde{U}(i, t) \}$ are the weight and choice variables for the smaller population, then $\tilde{U}(i, t) = U(i, t)$ and $\tilde{W}(i, j, t + 1) = W(i, j, t + 1)$ for all $t < \tau$ where τ is the first time, possibly infinite, at which $U(i, t)$ contains an edge between $[K]$ and $\{K + 1, \ldots, N\}$. In general, coupling methods show that if $\mathbb{P}(G_\infty = G_0 | t) > 1 - \epsilon$ then the conditional distribution of the Three's Company process from time t onward given F_t and $G_\infty = G_0$, shifted back t time units and restricted to a component H of G_∞, is within ϵ in total variation of the distribution of the Three's Company process on H started with initial weights $W'(i, j, 0) := W(i, j, t)$. $\square$

The Three's Company process on a population of size 3, 4, or 5 with any discount rate $1 - x < 1$ is ergodic: to see this just note that the Markov chain whose state space is the collection of W variables is Harris recurrent as a consequence of Lemma 7.2. The invariant measure gives positive weight to each edge, so each agent chooses each other with positive frequency, finishing the proof of Theorem 4.1.

Acknowledgment

Research supported in part by National Science Foundation grant # DMS 0103635.

References

Beggs, A. (2002) "On the convergence of reinforcement learning." Preprint Oxford Economics Discussion Paper 96, 1–34.

Benaïm, M. (1998) "Recursive algorithms, urn processes and chaining number of chain-recurrent sets." *Ergodic Theory and Dynamical Systems* 18: 53–87.

Benaïm, M. (1999) "Dynamics of Stochastic approximation algorithms." *Seminaires de Probabilités XXXIII*. Lecture Notes in Mathematics, vol. 1709: 1–68. Berlin: Springer.

Benaïm, M. and M. Hirsch (1995) "Dynamics of Morse–Smale urn processes." *Ergodic Theory and Dynamical Systems* 15: 1005–30.

Bereby-Meyer, Y. and I. Erev (1998) "On learning to become a successful loser: a comparison of alternative abstractions of learning processes in the loss domain." *Journal of Mathematical Psychology* 42: 266–86.

Bonacich, P. and T. Liggett (2003) "Asymptotics of a matrix-valued Markov chain arising in sociology." *Stochastic Processes and their Applications* 104: 155–71.

Busemeyer, J. and J. Stout (2002) "A contribution of cognitive decision models to clinical assessment: decomposing performance on the Bechara Gambling Task." *Psychological Assessment* 14: 253–62.

Busemeyer, J. and J. Townsend (1993) "Decision field theory: a dynamic–cognitive approach to decision making in an uncertain environment." *Psychological Review* 100: 432–59.

Bush, R. and F., Mosteller (1955) *Stochastic Models for Learning*. New York: Wiley.

Camerer, C. and T. Ho (1999) "Experience-weighted attraction in games." *Econometrica* 64: 827–74.

Coppersmith, D. and P. Diaconis (1997). Unpublished manuscript.

Davis, B. (1990) "Reinforced random walk." *Probability Theory and Related Fields* 84: 203–29.

Davis, B. (1999) "Reinforced and perturbed random walks. Random Walks." *Bolyai Society Mathematical Studies* 9: 113–26 (Budapest, 1998).

Eggenberger, F. and G. Pólya (1923) "Uber die Statistik verketter vorgäge." *Zeitschrift für Angewandte Mathematik und Mechanik* 1: 279–89.

Erev, I. and A. Roth (1998) "Predicting how people play games: reinforcement learning in experimental games with unique mixed-strategy equilibria." *American Economic Review* 88: 848–81.

Herrnstein, R. J. (1970) "On the law of effect." *Journal of the Experimental Analysis of Behavior* 13: 243–66.

Limic, V. (2001) "Attracting edge property for a class of reinforced random walks." Preprint. <http://www.math.cornell.edu/limic/>.

Luce, D. (1959) *Individual Choice Behavior*. New York: Wiley.

Pemantle, R. (1988) "Phase transition in reinforced random walk and RWRE on trees." *Annals of Probability* 16: 1229–41.

Pemantle, R. (1992) "Vertex-reinforced random walk." *Probability Theory and Related Fields* 92: 117–36.

Pemantle, R. and B. Skyrms (2003) "Reinforcement Schemes May Take a Long Time to Exhibit Limiting Behavior." In preparation.

Pemantle, R. and B. Skyrms (2004) "Time to absorption in discounted reinforcement models." *Stochastic Processes and their Applications* 109: 1–12.

Roth, A. and I. Erev (1995) "Learning in extensive form games: experimental data and simple dynamic models in the intermediate term." *Games and Economic Behavior* 8, 164–212.

Skyrms, B. and R. Pemantle (2000) "A dynamic model of social network formation." *Proceedings of the National Academy of Sciences of the United States of America* 97: 9340–6.

Suppes, P. and R. Atkinson (1960) *Markov Learning Models for Multiperson Interactions*. Palo Alto, CA: Stanford Univ. Press.

13

Time to Absorption in Discounted Reinforcement Models

with Robin Pemantle

1. Introduction

Stochastic models incorporating mechanisms by which likelihoods of outcomes increase according to their accumulated frequencies have been around since the introduction of Pólya's Urn (Eggenberger and Pólya 1923). The mathematical framework for many of these models appears in the literature on stochastic approximation, beginning with Robbins and Monro (1951), in the urn model literature (Freedman 1965), in the literature on reinforced random walks (Pemantle 1990; Limic 2003), and in the literature on the relation between stochastic systems and their deterministic mean-field dynamical system approximations (Benaïm and Hirsch 1995; Benaïm 1996).

These processes, known in the mathematical community as *reinforcement processes*, have long been used by psychologists as models for learning (Bush and Mosteller 1955; Iosifescu and Theodorescu 1969; Norman, 1972; Lakshmivarahan 1981). Increasingly, reinforcement models have been adopted by other social scientists as interactive models in which collective learning takes place in the form of network formation or adaptation of strategies: sociologists studying the "small world" network phenomenon (Watts and Strogatz 1998; Barrat and Weigt 2000), formation of dyads of reciprocal approval (Flache and Macy 1996), economists studying evolutionary game theory (Maynard Smith 1982), strategic learning (Fudenberg and Kreps 1993; Roth and Erev 1995) or its interaction with network structure (Ellison 1993; Anderlini and Ianni 1997). These models are designed to explore mechanisms by which agents with limited information, rationality, or sophistication may nevertheless achieve advantageous social structures via the application of simple rules for behavior change.

Due, perhaps, to a dearth of types of simple local rules (or perhaps to a lack of imagination on the part of modelers), most reinforcement models fall into one of two classes. The first class contains models for which the past is weighted uniformly. This class includes the urn models, stochastic approximations and reinforced random walks mentioned above, as well as the economic game theory models of Roth and Erev (1995). Uniform weighting means that the step from time $n - 1$ to n represents a fraction of only $1/n$ of the total learning up to time n, so one obtains a time-inhomogenous process in which the hidden variables change by amounts on the order of $1/n$ at time n. The second class consists of models in which the past is exponentially discounted or forgotten. This class includes the learning models of the 1960s and 1970s (Iosifescu and Theodorescu 1969), as well as many contemporary models of repeated economic games (e.g. Bonacich and Liggett 2002). In these models, the weight of the present is asymptotically equal to the discount parameter, x, defined as the x for which an action t units of time will be weighted by $(1 - x)^t$. More precisely, the fraction of total learning between time $n - 1$ and n of the total learning to time n will be roughly the maximum of $1/n$ and x.

Our chief concern in this chapter is to study how the discounted process approaches the non-discounted process as $x \to 0$. The long-run behavior in these two cases is qualitatively different. Limit theorems for non-discounted processes have been obtained chiefly in the framework of stochastic perturbations to dynamical systems. Typically, the stochastic system converges to limit points or limit cycles of the dynamical system that corresponds to the mean motion of the stochastic system (Benaïm and Hirsch 1995). The random limit is supported on weakly stable equilibria (Pemantle 1990), though the system may remain near unstable equilibria for long periods of time (see Benjamini and Pemantle 2003) for a discussion of this phenomenon in continuous time).

In the discounted processes we study here, there are trapping states, into which the chain must eventually fall (there is another kind of discounted process we are not concerned with here, which converges to an ergodic Markov chain, see, e.g., Iosifescu and Theodorescu 1969). The reciprocity model of Bonacich and Liggett (2002), for example, is of this type. From the point of view of studying the transition as discounting goes to zero, the most interesting case is when the trapping states are disjoint from the stable equilibria in the non-discounted process. Trapping states for the discounted process must always be equilibria for the non-discounted process, but when all the trapping states are unstable equilibria (equivalently, all the stable equilibria of the non-discounted process are non-trapping in the discounted process), then the conflict between the discounted and non-discounted behavior is maximized.

The transition is easy to describe informally. As the discount rate approaches zero, the discounted process behaves for a longer and longer time like the non-discounted process, and then abruptly falls into a trap. Of course, when the discounting parameter is x it takes time at least on the order of $1/x$ for the system to notice there is discounting going on. But in fact, due to the learning that has gone on during this phase, it will take time of order $\exp(cx^{-1})$ before the system discovers a trap and falls in. It is, in fact, not hard to guess this via back-of-the-napkin computations. One of our main motivations for pursuing this rigorously was to explain why simulation data contradicted the easily proved limit theorem: it was because the time scale of the simulation (let alone of any real phenomenon modeled by the simulation) was never anywhere near the time needed to find a trap.

Our purpose in the present paper is to prove various versions of this. In the next section, we present the ternary interaction model which was our original motivation for this study. Section 3 then introduces a simple process that is a building block for the ternary interaction model. For that process, results about trapping times can be proved with the correct constant. The last section then proves an $\exp(cx^{-1})$ waiting time result for a general class of models, but without the correct value of c. With a little linear algebra, this is shown to apply to the ternary interaction model of Section 2.

2. Three's Company: A Ternary Interaction Model

The following process is described in Pemantle and Skyrms (2003), where it is called Three's Company, and is put forth as a model for the formation of ternary collaborations in a three-player version of

Rousseau's Stag Hunting game. Fix a positive integer $N \geq 4$, representing the size of the population. For $t \geq 0$ and $1 \leq i, j \leq N$, define random variables $W(i, j, t)$ and $U(i, t)$ inductively on a common probability space $(\Omega, \mathscr{F}, \mathbb{P})$ as follows. The W variables are positive numbers, and the U variables are subsets of the population of cardinality 3. One may think of the U variables as random triangles in the complete graph with a vertex representing each agent. The initialization is $W(i, j, 0) = 1$ for all $i \neq j$, while $W(i, i, 0) = 0$). The inductive step, for $t \geq 0$, defines probabilities (formally, conditional probabilities given the past) for the variables $U(i, t)$ in terms of the variables $W(r, s, t)$, $r, s \leq N$, and then defines $W(i, j, t + 1)$ in terms of $W(i, j, t)$ and the variables $U(r, t)$, $r \leq N$. The equations are:

$$\mathbb{P}\Big(U(i,t) = S \mid \mathscr{F}_t\Big) = \frac{1_{i \in S}\, \Pi_{r,s \in S, r < s} W(r,s,t)}{\sum_{S': i \in S'}\, \Pi_{r,s \in S', r < s} W(r,s,t)}, \qquad (2.1)$$

$$W(i, j+1) = (1-x)W(i,j,t) + \sum_{r=1}^{N} 1_{i,j \in U(r,t)}. \qquad (2.2)$$

Here $(1 - x)$ is the factor per unit time by which the past is discounted, and the σ-field conditioned on is the process up to time t,

$$\mathscr{F}_t := \sigma\{W(i, j, u): u \leq t\}.$$

We may think of the normalized matrix

$$\mathbf{W}_t := \frac{1}{\sum_{i,j} W(i,j,t)} W(\cdot, \cdot, t)$$

as the state vector, which is then an asymptotically time-homogeneous Markov chain, with an evolution rule of the well-known form

$$E(\mathbf{W}_{t+1} - \mathbf{W}_t \mid \mathscr{F}_t) = g(t)[\mu(\mathbf{W}_t) + \xi_t], \qquad (2.3)$$

where in this case, $g(t) = 1/x + O(1/t)$, the drift vector field μ maps the simplex of normalized matrices into its tangent space and may be explicitly computed, and ξ_t are martingale increments of order 1.

These equations model a social interaction in which each agent i at each time t invites two others to frolic.[1] For each agent i, the trio chosen by i is chosen from all possible trios containing i, according to the

---

[1]   Engage in some rewarding interaction such as anti-competitive price fixing.

products of the weights $W(i, \cdot, t)$. Thus the probability of agent i forming the trio $\{i, j, k\}$ is proportional to $W(i, j, t)W(i, k, t)$. After the frolicking, fond memories ensue: each of the three pair weights $W(i, j, t)$, $W(i, k, t)$, and $W(j, k, t)$ is increased by 1, and a portion x of the past weights is forgotten. For ease of bookkeeping, the weights of unordered pairs are defined as symmetric weights of ordered pairs, so the weights $W(j, i, t)$, $W(k, i, t)$, and $W(k, j, t)$ are increased as well. We write $W(e, t)$ for $W(i, j, t)$ when e is the edge (unordered set) $\{i, j\}$.

It is shown in Pemantle and Skyrms (2003a) that the network always breaks into small cliques, with interactions occurring only among the cliques.

THEOREM 2.1 *In Three's Company, with population size $n \geq 4$ and any discount rate $x \in (0,1)$, with probability 1 the population may be partitioned into subsets of sizes 3–5, such that each member of each subset chooses each other with positive limiting frequency, and chooses members outside the subset only finitely often. Every partition into sets of sizes 3–5 has positive probability of occurring.*

Simulation data are also given there. For $N = 6$, if $x = 0.4$ (a rather steep discount rate), the network always breaks into two cliques of size 3, as predicted by the theorem. When $N = 6$ and $x = 0.2$, which is still a greater discount rate than one finds in most economic models, one finds, with runs of several thousand, that no such structure emerges. Instead, all six members of the population remain well connected. This is because of the exponential time scale of the transition from stable equilibria (well connectedness is a stable equilibrium of the non-discounted model) to trapping states (two cliques of size three is the unique trapping state of the $N = 6$ discounted model). Specifically, in the last section of this chapter we will prove.

THEOREM 2.2 *In the game Three's Company, for each $N \geq 6$ there is a $\delta > 0$ and numbers $c_N > 0$ such that in Three's Company with N players and discount rate $1-x$, the probability is at least δ that each player will play with each other player beyond time $\exp(c_N x^{-1})$.*

3. Trapping in One Dimensional Discounted Reinforcement

In this section we analyze a one dimensional process in which sharp quantitative results may be obtained on the exponential rate at which the time until trapping increases with $1/x$, where $1-x$ is the discount factor. This is in keeping with our philosophy of providing sharp results on a collection of simplified models that constitute building blocks for more complicated models. In the last section we will apply the principles gleaned from this to get bounds on the exponential rate of increase of trapping time in the Three's Company model.

Let us consider a system whose state vector varies in the interval $[0, 1]$ with evolution dynamics that are symmetric around an attractor at $1/2$, and whose transitions from state w have a profile that depends on w and is scaled by x. In analogy with models such as Three's Company, we assume that the unscaled transitions have variance bounded from below as w varies over compact sub-intervals of $(0, 1)$. Thus the rules of evolution of the state vector W may be given in terms of probability distributions Q_w, parametrized by $w \in [0, 1]$, with bounded support, satisfying $Q_w(s) = Q_{1-w}(-s)$, and obeying

$$\mathbb{P}\left(W(n+1) - W(n) \in x \cdot S \mid \mathscr{F}_n\right) = Q_{W(n)}(S) \qquad (3.1)$$

on the event that $W(n)$ is in a compact subinterval I_x, with $I_x \uparrow (0, 1)$ as $x \to 0$. We assume that the mean of Q_w is positive on $(0, 1/2)$ and negative on $(1/2, 1)$, but that Q_w has both positive and negative elements in its support and varies smoothly with w.

As an example, one may consider a class of two-color urn models generalizing Freedman's (1949, 1965) Urn in the discounted setting. An urn begins with $R(0)$ red balls and $B(0)$ black balls. At the nth time step, a random number $U(n)$ of red balls and $V(n)$ black balls are added to the urn. Conditional on the past, $\mathbb{P}(U(n) = k) = u_{W(n-1)}(k)$ and $\mathbb{P}(V(n) = k) = u_{1-W(n-1)}(k)$, where $W(n) := R(n)/(R(n) + B(n))$ is the state parameter, in this case the proportion of red balls, and u_w are probability distributions on the nonnegative integers, continuously varying in the parameter $w \in [0, 1]$, satisfying

$$\frac{\sum_k k u_w(k)}{\sum_k k\left(u_w(k) + u_{1-w}(k)\right)} > w \quad \text{for } 0 < w < \frac{1}{2}.$$

At the end of each step, all balls are reduced in weight by a factor of $1-x$. For greater specificity, one may keep in mind an example where two balls are sampled: if they are of the same color then one ball of that color is added; if they are of different colors then one ball of each color is added.

In the non-discounted system, where the step size scales as $1/n$ at time n instead of holding constant at x, the system is well approximated by a diffusion with incremental variance of order n^{-2} and drift $n^{-1}\mu_w$, with μ_w being the mean, $\overline{Q_w}$, of Q_w. Thus (2.3) holds with $g(t) = t^{-1}$ and $u(w) = \overline{Q_w}$. The system must converge to the unique attracting equilibrium at $1/2$ (Pemantle, 1990). In the discounted case, although the state must converge to 0 or 1, the logarithm of the expected time to come near 0 or 1 may be computed in terms of the following data.

Pick any $w \in (0, 1/2)$. The quantity

$$Z_w(\lambda) := \int \exp(-\lambda y) \, dQ_w(y)$$

is equal to 1 at $\lambda = 0$. The derivative $(d/d\lambda)Z_w(\lambda)|_{\lambda = 0}$ is given by $\int (-y) \, dQ_w(y)$, which is negative by the assumption that Q_w has positive mean. On the other hand, since Q_w gives positive probability to negative values, we see that as $\lambda \to \infty$, $Z_w \to \infty$, and by convexity of $Z_w(\cdot)$ it follows that there is a unique $\lambda_w > 0$ for which $Z_w(\lambda_w) = 1$. Define

$$\Lambda(w) := \int_w^{1/2} \lambda_u \, du$$

and let $C := \Lambda(0)$.

THEOREM 3.1 Let $I_x \uparrow (0, 1)$ as $1-x \uparrow 1$, *slowly enough so that transitions outside of* $[0, 1]$ *are never possible. Let T_x be the expectation of the first time n that $W(n) \notin I_x$. Then as $1-x \uparrow 1$,*

$$x \log \mathbb{E}T_x \to C.$$

REMARK This is essentially a large deviation problem, so the rate C is not determined by the mean and variance of Q_w but rather by the exponential moments of Q_w. In particular, there are many processes which satisfy (2.3) with the same g and μ, but their large deviation rates depend on the fine structure of the increment distribution through the exponential moments, as captured by Z_w and λ_w. The solution of this rate problem is standard; a similar analysis may be found, for example in Dembo and Zeitouni (1993: sect. 5.8.2).

Proof: For one inequality, we fix any $\delta > 0$. Define the quantity

$$M_{(\delta)}(t) := \exp\big((1 - \delta)x^{-1}\Lambda\big(W(t)\big)\big).$$

Since Q_w varies smoothly with w and has bounded support, we see that as $x \to 0$,

$$\Lambda(W(t+1)) - \Lambda(W(t)) = (W(t+1) - W(t))(-\lambda_{W(t)} + O(x)).$$

Therefore, since the conditional distribution of $x^{-1}(W(t+1) - W(t))$ given $\mathscr{F}_t$ is given by $Q_{W(t)}$, we see that

$$
\begin{aligned}
\mathbb{E}(M_{(\delta)}(t+1)|\mathscr{F}_t) &= M_{(\delta)}(t)\mathbb{E}[\exp((1-\delta)x^{-1}(W(t+1) \\
&\quad -W(t))(-\lambda_{W(t)} + O(x)))] \\
&\to M_{(\delta)}(t)Z_{W(t)}((1-\delta)\lambda_{W(t)})
\end{aligned}
\tag{3.2}
$$

uniformly in $W(t)$ as $x \to 0$. We know that $Z_w(\cdot) < 1$ on $(0, \lambda_w)$, hence we may pick $x = x(\delta)$ small enough so that

$$M_{(\delta)}(t)^{-1}\mathbb{E}(M_{(\delta)}(t+1)|\mathscr{F}_t) < 1$$

or in other words, so that $M_{(\delta)}$ is a supermartingale.

Let $I_x = [a_x, 1-a_x]$. Starting with $W(t_0) \in (1/2-\delta, 1/2)$, and stopping at the time τ when $W(\cdot)$ exits $[a_x, 1/2]$, we have for some constant $c(\delta)$ going to zero with δ,

$$
\begin{aligned}
\exp(x^{-1}c(\delta)) &\geq M_{(\delta)}(t_0) \\
&\geq \mathbb{E}(M_{(\delta)}(\tau)|\mathscr{F}_{t_0}) \\
&\geq \mathbb{P}(M_{(\delta)}(\tau) < a_x)\exp(x^{-1}\Lambda(a_x))
\end{aligned}
$$

which implies that

$$\log\mathbb{P}(M_{(\delta)}(\tau) < a_x) < -(1-\delta)x^{-1}(\Lambda(0) + o(1))$$

as $x \to 0$. A completely analogous argument shows that the process started in $(1/2, 1/2 + \delta)$ exits $[1/2, 1 - a_x]$ with at most this probability as well. The trajectory of $W(\cdot)$ may be decomposed into segments that begin in $[1/2 - \delta, 1/2 + \delta]$ and end when $W(t) - 1/2$ changes sign or $W(t) \notin I_x$. We have shown that the expected number of trajectories is at least $\exp(x^{-1}(1 - \delta)(\Lambda(0) + o(1)))$ as $x \to 0$, which implies that the number of time steps until exiting I_x is at least this great, once $W(t) \in [1/2 - \delta, 1/2 + \delta]$. Letting $c'(\delta)$ denote the probability of entering this interval, we see that

$$\mathbb{E}T_x \geq c'(\delta)\exp(x^{-1}(1 - \delta)(\Lambda(0)o(1)))$$

as $x \to 0$, and finally, sending δ to zero proves that

$$\liminf x \log \mathbb{E}T_x \geq C.$$

For the other direction, define a tilted measure on the space of trajectories $\{W(t): t = 0,1,2,\dots\}$ as follows. Eq. (3.1) is replaced by

$$\tilde{\mathbb{P}}(W(n+1) - W(n) \in x \cdot S \,|\, \mathscr{F}_n) = \tilde{Q}_{W(n)}(S),$$

where $\delta > 0$ is fixed and the Radon–Nikodym derivative is given by

$$\frac{d\tilde{Q}_w}{dQ_w}(y) = \frac{\exp[(1+\delta)(\Lambda(w+y) - \Lambda(w))]}{\int \exp[(1+\delta)(\Lambda(w+y) - \Lambda(w))]dQ_w(y)}.$$

The measure $\tilde{\mathbb{P}}$ is designed to have two properties. First, the process $\{W(t)\}$ is a supermartingale on $[a_x, 1/2]$ with respect to $\tilde{\mathbb{P}}$ for sufficiently small x. To see this, note that this is equivalent to $\tilde{Q}_w$ having negative mean, which is equivalent to

$$\int y e^{(1+\delta)(\Lambda(w+y)-\Lambda(w))} dQ_w(y) \leq 0$$

for all $w \in [a_x, 1/2]$. The quantity $\Lambda(w + y) - \Lambda(w)$ is equal to $y(-\lambda_w + o(1))$ as $x \to 0$, so it suffices to show that

$$\int y e^{-(1+\delta)y\lambda_w} dQ_w(y) < 0.$$

But this follows from the fact that Z_w is convex and increases through 1 at λ_w: the derivative at $(1 + \delta)\lambda_w$ must therefore be positive, and the derivative may be identified as

$$-\int y e^{-(1+\delta)\lambda_w y} dQ_w(y),$$

proving the supermartingale property.

The second property is that if τ is the exit time of $[a_x, 1/2]$, then on the σ-field $\mathscr{F}_\tau$, $d\tilde{\mathbb{P}}/d\mathbb{P}$ is at most $\exp((1 + \delta)x^{-1} \Lambda(0))$. Indeed, by its definition,

$$\frac{d\tilde{\mathbb{P}}}{d\mathbb{P}}(W(t_0),\dots,W(\tau)) = \prod_{t=t_0}^{\tau-1} \frac{\exp[x^{-1}(1+\delta)(\Lambda(W(t+1) - \Lambda(W(t))))]}{\int \exp[(1+\delta)(\Lambda(W(t) + y) - \Lambda(W(t)))]dQ_{W(t)}(y)}.$$

The denominator of each factor is at least 1 by the fact that

$$M_{(-\delta)}(t) := \exp((1 + \delta)x^{-1}\Lambda(W(t)))$$

is a submartingale when $x(\delta)$ is small enough, which is proved by a computation exactly analogous to (3.2). The product of the numerators is simply $\exp((1 + \delta)x^{-1} \Lambda(W(\tau)) - \Lambda(W(t_0)))$, which is at most $\exp((1 + \delta)x^{-1} \Lambda(0))$, proving the second property.

Running $\tilde{\mathbb{P}}$ on $[a_x, 1/2]$ and its reflection on $[1/2, 1-a_x]$, the process $1/2 - |1/2 - W(t)|$ is a supermartingale with incremental variance of order x^{-2}. The median time for it to reach a value less than a_x is therefore at most $O(x^{-2})$. Comparing $\tilde{\mathbb{P}}$ and $\mathbb{P}$, we find that there is a c such that from any starting data, the probability of exiting I_x by time cx^{-2} is at least. $(1/2)(d\mathbb{P}/d\tilde{\mathbb{P}}) \geq (1/2)\exp(-(1 + \delta)x^{-1}\Lambda(0))$. Breaking into time intervals of size cx^{-2}, it then follows that the mean time to exit I_x is at most $2cx^{-2} \exp(C(1 + \delta)x^{-1})$. As this holds for any $\delta > 0$ (and constants depending on δ), this proves that

$$\limsup x \log \mathbb{E} T_x \leq C$$

and finishes the proof of the theorem. $\square$

4. Proof of Theorem 2.2

In analogy with the one-dimensional toy model, we expect to find an exponential wait to trapping in Three's Company if the non-discounted system has an attractor outside of the limit set of absorbing states of the discounted system. Unfortunately, at this point we cannot see any way to compute the large deviation rate in multi-dimensional problems. The standard multi-dimensional analogue to Theorem 3.1 is expressed as a variational result involving minimizing a functional over all paths. We settle for proving the existence of a non-zero exponential rate in x^{-1}. The following result will imply Theorem 2.2.

PROPOSITION 4.1 *Let the vector-valued Markov chain $\vec{W}(t)$ satisfy*

$$\mathbb{P} \, \vec{W}((t+1) - \vec{W}(t) \in xS \mid \vec{W}(t)) = Q_{\vec{w}(t)}(S) \qquad (4.1)$$

with Q_w having bounded support and varying smoothly as w varies over some closed neighborhood $\mathcal{N}$ of a point $\vec{c}$. Suppose there is a strong Liapunov function V, meaning that V is smooth and bounded with $V < 0$ on $\mathcal{N}^c$, $V(\vec{c}) > 0$ and

$$\int V(w+y) \mathrm{d}Q_w(y) > V(w)$$

for all $w \in \mathcal{N}$. *Then there is a constant* γ *such that for all* $\vec{h}$ *in some smaller neighborhood of* $\vec{c}$,

$$\mathbb{E}_{\vec{h}} T_x > \gamma \, \exp(\gamma x^{-1})$$

for sufficiently small x, *with* T_x *being the time to exit* $\mathcal{N}$.

Proof Given a non-negative parameter, λ, define $M(t) = M_\lambda(t) = \exp(-\lambda V(\vec{w}(t)))$. Arguing as in the first half of the proof of Theorem 3.1, we see from the bounded support hypothesis that for fixed $w \in \mathcal{N}$,

$$M_\lambda(t)^{-1} \mathbb{E}(M_\lambda(t+1) - M_\lambda(t) | \mathscr{F}_t) \tag{4.2}$$

vanishes at $\lambda = 0$ and has negative derivative. By compactness of $\mathcal{N}$ and smoothness of Q_w, we may choose a $\lambda > 0$ so that (4.2) is negative for all $w \in \mathcal{N}$, implying that M(t) is a supermartingale up to the exit time of $\mathcal{N}$. Let $\mathcal{N}'$ be the neighborhood $\{\vec{h} \in \mathcal{N} : V(h) > V(\vec{c})/2\}$. Using T_G to denote the first time $\tau \geq 0$ that $W(\tau) \in G$, we then have, for $V(\vec{h}) > V(\vec{c})/4$,

$$
\begin{aligned}
e^{-\lambda V(\vec{c})/4} &\geq \mathbb{E}_{\vec{h}}(\vec{W}(0)) \\
&\geq \mathbb{E}_{\vec{h}} T_{\mathcal{N}^c \cup \mathcal{N}'} \\
&\geq \mathrm{P}_{\vec{h}}(T_{\mathcal{N}^c} < T_{\mathcal{N}^c})
\end{aligned}
$$

Breaking the time $T_{\mathcal{N}^c}$ into sojourns away from $\mathcal{N}'$ then proves the theorem with $\gamma = \lambda V(\vec{c})/4$.

In Three's Company, if we start with the sum of the weights equal to $3Nx^{-1}$ (that is, in stationarity), then the dynamics are described exactly by (4.1). We need only check the existence of a strong Liapunov function. This will follow if we can identify a hyperbolic attractor for the vector field $F(\cdot)$, where $F(w)$ is the mean of Q_w. Indeed, if F vanishes at a point $\vec{c}$ and $dF(\vec{c})$ has eigenvalues with negative real parts, then there is a quadratic function V near $\vec{c}$ satisfying $\bigtriangledown V \cdot F < 0$ which we may take as the Liapunov function. All that remains is to identify the hyperbolic attractor for the mean motion field.

The mean motion is given by a vector field F on the state space. The state space is the set of non-negative real functions X on the edges summing to $3Nx^{-1}$, which we think of as embedded in the cone of nonnegative functions, since F extends naturally via $F(\lambda X) = F(X)$. The computations are a little more convenient when we normalize the sum of weights to be $\binom{N}{2}$. It is also convenient to let $n = N-1$ be one less than the number of agents. The attractor on which we focus is the symmetric point c defined by $c(e) = 1$ for all e. It is immediate to verify that $F(\vec{c}) = 0$. In order to verify that $\vec{c}$ is an attractor for F, we need to compute the differential of F at $\vec{c}$. Accordingly, let 1_e denote the function that is 1 on e and 0 elsewhere. The derivative of F in the 1_e direction is computed as follows.

Let the edge weights at time t be given by $1 + \varepsilon 1_e$. The expected number of i for which $f \in U(i, t)$ is $6/n + O(\varepsilon)$ for all f. By symmetry, the $O(\varepsilon)$ term depends only on whether f shares two, one, or zero endpoints with e. For example, in the case $f = e$, we compute the expected number of times e is reinforced as follows. Let $e = \{v, w\}$. Then

$$\mathbb{P}(e \in U(v, t)) = (n - 1) \frac{1 + \varepsilon}{\binom{n}{2} + (n - 1)\varepsilon}$$

The probability of $e \in U(w, t)$ is the same. For $z \neq v, w$, the probability of $e \in U(z, t)$ is exactly $\binom{n}{2}^{-1}$. Summing yields an expected increment in $W(e, t)$ of

$$2(n - 1) \frac{1 + \varepsilon}{\binom{n}{2} + (n - 1)\varepsilon} + \frac{n - 1}{\binom{n}{2}} = \frac{6}{n} + \frac{4(n - 2)}{n^2}\varepsilon + O(\varepsilon)^2.$$

We write this as

$$\frac{6}{n}\left(1 + \frac{2(n - 2)}{3n}\varepsilon + O(\varepsilon^2)\right).$$

Computing, the other two expectations in this manner, we find that the expectation for f at distance j from e is $(6/n)(1 + B_j \varepsilon + O(\varepsilon^2))$, where

$$B_0 = \frac{2(n-2)}{3n},$$

$$B_1 = 0,$$

$$B_2 = -\frac{4}{3n(n-1)}.$$

From this it follows that for any $\vec{h}$,

$$\frac{n}{6}\mathbb{E}(W(t+1) \mid W(t) = \vec{c} + \varepsilon\vec{h}) = \vec{c} + M \to h + \mathrm{O}(|\vec{h}|^2).$$

where M is a generalized circulant matrix (symmetric under the action of edge permutation on pairs of edges) with entries B_0 on the diagonal, B_2 for disjoint edges, and zero otherwise. Since F is the vector field pointing toward $\mathbb{E}(W(t+1) - W(t)|W(t))$, the differential of F is, up to the constant multiple $6/n$, equal to $M - I$.

The eigenvalues of a matrix such as M are particularly easy to evaluate, using the rubric of *association schemes* (see Terwilliger, 1996: sect. 2.2, which is taken from Bannai and Ito, 1984; Brouwer et al., 1989). All such matrices are elements of the Bose–Mesener algebra $\mathcal{M}$ which, in the case of the incidence graph for edges of the complete graph, is commutative semi-simple of dimension 3. This implies that M has at most three distinct eigenvalues. These may be found by computing the action of M on the three shared eigenspaces common to all elements of $\mathcal{M}$.

The null eigenspace has dimension 1: $M\vec{h} = 0$ if and only if $\vec{h} = \lambda\vec{c}$. The other two eigenvalues may be gotten by choosing an edge e and setting the eigenvectors equal to $a_2 H_2 + a_1 H_1 + a_0 H_0$, where $H_2 = \mathbf{1}_e$, H_1 is the sum of $\mathbf{1}_f$ over edges f sharing one vertex with e, and H_0 is the sum of $\mathbf{1}_f$ over edges f disjoint from e. The action of M on such a sum produces another such sum, and is linear, having matrix

$$\frac{4}{3n(n-1)} \begin{pmatrix} \binom{n-1}{2} & 0 & -1 \\ 0 & \binom{n-2}{2} & -2(n-3) \\ -\binom{n-1}{2} & -\binom{n-2}{2} & 2n-5 \end{pmatrix}$$

with respect to a, b, and c. The left eigenvectors of this are

$$(1,1,1), \quad \left(\frac{n-1}{2},\frac{n-3}{4},-1\right) \quad \text{and} \quad \left(\binom{n-1}{2},-\frac{n-2}{2},1\right).$$

The corresponding eigenvalues are

$$0, \quad \frac{2}{3}\frac{(n+1)(n-2)}{n(n-1)} \quad \text{and} \quad \frac{2n-3}{3n-1}.$$

The equations of mean motion are

$$\mathbb{E}(W(t+1)-W(t)\mid W(t)-\vec{c}+\varepsilon\vec{h}) = \frac{6}{n}x^{-1}(M-I)\vec{h}+\mathrm{O}(|\vec{h}|^2),$$

whence the point $\vec{c}$ is attracting for sufficiently small x if and only if the real parts of all eigenvalues of M are less than 1. We have identified that this is so, and therefore $\vec{c}$ is attracting. The hypotheses of Proposition 4.1 are therefore satisfied with a quadratic Liapunov function, and Theorem 2.2 follows from Theorem 3.1.

References

Anderlini, L. and A. Ianni (1997) "Learning on a torus." In *The Dynamics of Norms* ed. by C. Bicchieri, R. Jeffrey, B. Skyrms, 87–107. Cambridge: Cambridge University Press.

Bannai, E. and T. Ito (1984) *Algebraic Combinatorics I: Association Schemes.* Menlo Park, CA: Benjamin/Cummings.

Barrat, A. and M. Weigt (2000) "On the properties of small-world network models." *European Physical Journal* B 13: 547.

Benaïm, M. (1996) "A dynamical system approach to stochastic approximations." *SIAM Journal Control and Optimization* 34: 437–72.

Benaïm, M. and M. Hirsch (1995) "Dynamics of Morse–Smale urn processes." *Ergodic Theory Dynamical Systems* 15: 1005–30.

Benjamini, I. and R. Pemantle (2003) "Probabilities for cooled Brownian motion to linger near the top of a hill, and application to a market share model." Preprint.

Bonacich, P. and T. Liggett (2002) "Asymptotics of a matrix-valued Markov chain arising in sociology." Preprint.

Brouwer, A., A. Cohen, and A. Neumaier (1989) *Distance-Regular Graphs.* Berlin: Springer.

Bush, R. and F. Mosteller (1955) *Stochastic Models for Learning.* New York: Wiley.

Davis, B. (1990) "Reinforced random walk." *Probability and Theory Related Fields* 84: 203–29.

Dembo, A. and O. Zeitouni (1993) *Large Deviations Techniques and Applications.* Boston, MA: Jones and Bartlett.

Eggenberger, F. and G. Pólya (1923) "Uber die Statistik verketter vorgäge." *Zeitschrift für Angewandte Mathematik und Mechanik* 1: 279–89.

Ellison, G. (1993) "Learning, local interaction, and coordination." *Econometrica* 61: 1047–71.

Flache, A. and M. Macy (1996) "The weakness of strong ties: collective action failure in a highly cohesive group." *Journal of Mathematical Sociology* 21: 3–28.

Freedman, B. (1949) "A simple urn model." *Communications on Pure and Applied Mathematics* 2: 59–70.

Freedman, D. (1965) "Bernard Friedman's urn." *Annals of Mathematical Statistics* 36: 956–70.

Fudenberg, D. and K. Kreps (1993) "Learning mixed equilibria." *Games and Economic Behavior* 5: 320–67.

Iosifescu, M. and R. Theodorescu (1969) *Random Processes and Learning.* New York: Springer.

Lakshmivarahan, S. (1981) *Learning Algorithms: Theory and Applications.* New York: Springer.

Limic, V. (2003) "Attracting edge property for a class of reinforced random walks." *Annals of Probability* 31: 1615–54.

Maynard Smith, J. (1982) *Evolution and the Theory of Games.* Cambridge: Cambridge University Press.

Norman, M. (1972) *Markov Processes and Learning Models.* New York: Academic Press.

Pemantle, R. (1990) "Nonconvergence to unstable points in urn models and stochastic approximations." *Annals of Probability* 18: 698–712.

Pemantle, R. and B. Skyrms (2003) "Network formation by reinforcement learning: the long and medium run." *Mathematical Social Sciences* 48: 315–27.

Robbins, H. and S. Monro (1951) "A stochastic approximation method." *Annals of Mathematical Statistics* 22: 400–7.

Roth, A. and I. Erev (1995) "Learning in extensive form games: experimental data and simple dynamic models in the intermediate term." *Games and Economic Behavior* 8: 164–212.

Terwilliger, P. (1996) Algebraic Combinatorics. Unpublished lecture notes.

Watts, D. and S. Strogatz (1998) "Collective dynamics of 'small-world' networks." *Nature* 393: 440–2.

PART IV

Dynamics of Signals

Introduction

The topic of the first paper in this section is the spontaneous emergence of meaningful signaling between two agents using reinforcement learning. The setting is the simplest signaling game, of the type introduced by David Lewis in his book *Convention*. Lewis identified signaling systems, which transmit information perfectly, as strict equilibria of signaling games. But he left open the question of how and when we get to such an equilibrium when we start with no pre-existing meaning and no inclination towards one signaling system or another. Simulations had shown that it is possible to learn to signal, but this paper proves that in this game simple reinforcement learning converges to a signaling system with probability one. (The mathematics are all due to Argiento, Pemantle, and Volkov.)

If we start with no initial bias towards one or the other, each of the two signaling conventions emerges with probability ½. Things change in more general signaling games with N states, M signals, and N acts (Hu, Tarrés, and Skyrms). Learning to signal optimally is still possible, but no longer guaranteed.

For economics laboratory experiments close to the game analyzed here, see Blume et al. (1998, 2002). For theoretical analysis of infinite population evolutionary dynamics (replicator dynamics with and without mutation) applied to a 2 state, 2 signal, 2 act Lewis signaling game see Hofbauer and Huttegger (2008). Learning to signal by interaction with neighbors is analyzed by Zollman (2005) for a lattice, and Wagner (2009) for an arbitrary network.

The second selection modifies the basic reinforcement dynamics to allow the invention of new signals. In this way, the Lewis signaling game itself can evolve. It is possible to start the model with no signals at all, and allow the invention of signaling in an even stronger sense than before. The treatment here is partly analytic (by Zabell) and partly by simulation. Learning with invention appears to be remarkably effective in delivering optimal signaling systems, even settings where straight reinforcement may fail. A modification (by Alexander) introduces forgetting, which prunes excess signals.

The third and fourth essays combine signals with other games. Players exchange costless pre-play signals before they play a base game. In the third paper the analysis uses the replicator dynamics. It is shown that costless signals can change the stability properties of equilibria, create new equilibria, and dramatically change the size of basins of attraction. The measure of information in signals that I later use in my book, *Signals*, is introduced in this paper. Information in signals co-evolves with strategies in the base game. Transient information is important in changing the size of basins of attraction. In symmetric bargaining, pre-play signaling is conducive to evolution of the equal split. In the Stag Hunt game, pre-play signaling favors cooperation.

The following paper, with Francisco Santos and Jorge Pacheco, is also about pre-play signaling. It differs from the preceding in two major ways. First, it uses finite population dynamics, in which my co-authors are experts. Second, it varies the number of pre-play signals, where the previous paper used only two for the Stag Hunt. For the Stag Hunt, it turns out that more signals are better. With enough signals we spend almost all our time cooperating. Remarkably, many pre-play signals have the same sort of effect in Prisoner's Dilemma games. With enough "secret handshakes" available cooperators can stay ahead of defectors in the evolutionary race, and a high level of cooperation can be sustained. For effective pre-play signals in a different setting—local interaction with neighbors—see the second part of Zollman (2005).

The last paper deals with generalizations of the sender–receiver framework to one in which there are multiple senders and receivers. This takes a small step towards a more general theory of signaling networks. Receivers may get multiple signals, each carrying their own information. They may need to process that information in various ways in order to act optimally. Senders can send signals to many receivers in order to coordinate some task that requires teamwork. For some recent models with multiple senders see Barrett (2009, 2010, 2013), Godfrey-Smith (2013), and Wagner (2013b).

Evolutionary dynamics of costly signaling games have recently been explored by Huttegger and Zollman (2010) for the Sir Philip Sydney game in evolutionary biology, and by Wagner (2013a) for the Spence signaling game in economics. Godfrey-Smith and Martinez (2013) explore the relation between degree of common interest and successful signaling by systematic computer search. Wagner (2012) shows the existence of chaotic evolutionary dynamics in a signaling game. O'Connor (2013) shows how modified reinforcement learning with stimulus generalization leads to vague categories.

Returning to the concerns of the last section, suppose players have learned how to signal. How might they then self-assemble into signaling networks? Some beginning approaches, using low rationality learning, are in Huttegger, Skyrms, and Zollman (2013) and references cited there.

References

Barrett, J. (2009) "The Evolution of Coding in Signaling Games." *Theory and Decision* 67: 223–7.

Barrett, J. (2010) "Faithful Description and the Incommensurability of Evolved Languages." *Philosophical Studies* 147: 123–37.

Barrett, J. (2013) "On the Coevolution of Basic Arithmetic Language and Knowledge." *Erkenntnis* 78: 1025–36.

Barrett, J. and K. Zollman (2009) "The Role of Forgetting in the Evolution of Learning and Language." *Journal of Experimental and Theoretical Artificial Intelligence* 21: 293–309.

Blume, A., D. V. De Jong, Y.-K. Kim, and G. B. Sprinkle (1998) "Experimental Evidence on the Evolution of Meaning of Messages in Sender–Receiver Games." *The American Economic Review* 88: 1323–40.

Blume, A., D. V. De Jong, G. R. Neumann and N. E. Savin (2002) "Learning and Communication in Sender–Receiver Games: An Econometric Investigation." *Journal of Applied Econometrics* 17: 225–47.

Hofbauer, J. and S. Huttegger (2008) "Feasibility of Communication in Binary Signaling Games." *Journal of Theoretical Biology* 254: 843–9.

Godfrey-Smith, P. (2013) "Sender–Receiver Systems within and between Organisms." (Paper read at PSA 2012) *PhilSci Archive* University of Pittsburgh.

Godfrey-Smith, P. and M. Martinez (2013) "Communication and Common Interest." *PLOS Biology* DOI: 10.1371/journal.pcbi.1003282.

Hu, Y., P. Tarrés, and B. Skyrms (2011) "Reinforcement Learning in a Signaling Game." ArXiV, Preprint.

Huttegger, S., B. Skyrms, and K. Zollman (2013) "Probe and Adjust in Information Transfer Games." *Erkenntnis* doi: 10.1007/s10670-013-9467-y.

Huttegger, S. and K. Zollman (2010) "Dynamic Stability and Basins of Attraction in the Sir Philip Sydney Game." *Proceedings of the Royal Society B* 1925–32.

Mühlenbernd, R. (2011) "Learning with Neighbors." *Synthese* 183: 87–109.

Mühlenbernd, R. and M. Franke (2012) "Signaling Conventions: Who Learns What and Where in a Social Network." Preprint.

O'Connor, C. (2013) "The Evolution of Vagueness." *Erkenntnis* doi:10.1007/s10670-013-9463-2.

Wagner, E. (2009) "Communication and Structured Correlation." *Erkenntnis* 71: 377–93.

Wagner, E. (2012) "Deterministic Chaos and the Evolution of Meaning." *British Journal for the Philosophy of Science* 63: 547–75.

Wagner, E. (2013a) "The Dynamics of Costly Signaling." *Games* 4: 161–83.

Wagner, E. (2013b) "Divergent Interests and the Evolution of Inference." Working paper, University of Amsterdam.

Zollman, K. (2005) "Talking to Neighbors: The Evolution of Regional Meaning." *Philosophy of Science* 72: 69–85.

14

Learning to Signal: Analysis of a Micro-level Reinforcement Model

with Raffaele Argiento, Robin Pemantle, and Stanislav Volkov

1. Introduction

1.1 Motivation for the model

In recent decades, much attention has been given to repeated two-player, non-zero sum games and the evolution of strategy. The evolutionary game theory paradigm, originating in the late 1970s in works such as Taylor and Jonker (1978) has been thoroughly explored in a variety of contexts, with particular emphasis on explaining how cooperation might arise in games such as the Prisoner's Dilemma for which there seems to be an inherent disincentive to cooperate.

Another recent line of inquiry has been the formation of reasonable strategies within a population under myopic, *bounded rationality* types of constraints. The emphasis here is on finding evolutionary pathways whose mechanisms are simple enough that they may be employed by unsophisticated individuals, without much conscious thought, across a wide variety of contexts. This kind of mechanism can, among other things, hope to explain the formation of social and moral norms (Skyrms 2004; Alexander 2007). These norms are heuristics that may be easily understood and employed in a wide variety of contexts, and their formation is possible if it results from individual-level mechanisms that are advantageous when averaged over the many contexts in which they are employed.

The emphasis on the simplicity of micro-level mechanism has advantages beyond generalizability across contexts and levels of intelligence or conscious thought. It makes models more mathematically tractable. It also allows complexity to be increased in other dimensions, such as allowing the simultaneous evolution of strategy and network structure (Skyrms 2000, 2004). Keeping the number of parameters to a minimum allows, in principle, for empirical testing and calibration of the micro-level parameters (see, e.g. the discussion of the discount rate parameter in Pemantle and Skyrms 2003). On a theoretical level, finding the simplest, most parsimonious model to explain a phenomenon is generally thought to be a useful step in the investigation of the phenomenon in question.

The notion of an individual employing an urn model to govern plays in a repeated game is very old. The "two-armed bandit" problem, for instance (Bradt et al. 1956), features an individual trying to discover the state of nature and balance the considerations of optimal play under known information against play which will be most informative and thereby lead to gains in the future. One well-known strategy for this takes the form of an urn model; see for example Duflo (1996). This has been applied to commonly used protocols such as sequential sampling in medical trials (Wei and Durham 1978). Urn models are natural in the context of learning models for several reasons. It is not hard to posit micro-level psychological processes (such as extrapolating from available memories) that correspond well to the model. Urn models typically contain enough noise to avoid certain game-theoretically unstable equilibria while still possessing good convergence properties. Models that are neither quickly fixating (overly long memory) nor ergodic (overly short memory) correspond best to many qualitative learning phenomena.

Early instances of urn models arose as attempts to formulate reasonable strategies for one agent playing against Nature. More recent is the use of urns to model multi-player games in which simplicity at the micro-level is desired. A considerable number of formal interaction models have appeared in the last ten years in the fields of psychology, sociology, and political science. One may find many of these, for insance, in the literature on *Agent-based modeling*. This refers to a much broader class of formal systems that includes urn models; for examples of agent-based urn models in sociology, see Bonacich and Liggett (2030) or Flache and Macy (2002).

An important precursor to the agent-based modeling paradigm is the evolutionary game theory paradigm, which improves the explanatory power of classical game theory by incorporating a Darwinian population dynamic along with the strategic interaction. A more recent twist, introduced in Skyrms and Pemantle (2000), is to allow the network of interactions to evolve as well. The explanatory power of such systems and the philosophical ramifications of this are discussed in Skyrms (2004). Types of classical games to which such an analysis has been applied include Prisoner's Dilemmas, cooperative games in the vein of Rousseau's Stag Hunt, and bargaining games. In the present work, we take up the application of urn models to signaling games. The model we analyze here is the first in a series of models that incoporate successively more features of signaling problems. The more complex models are described briefly in the final section.

1.2 A two-state, two-signal communication game

We consider the following game, in which the players are the Sender, the Receiver, and Nature. Nature plays first, choosing a state of nature which is either 1 or 2. The Sender sees this play and must choose a signal. In this simplest nontrivial model, there are only two legal signals, A and B. The last play belongs to the Receiver, who sees the Sender's play but not Nature's play. The Receiver chooses an action from the set {1, 2}. The mutual goal is to have the action match the state of nature. The game is completely cooperative, in the sense that either both players win or both players lose. The game tree, shown in Figure 14.1, has eight paths, which we may denote by 1A1, 1A2, 1B1, 1B2, 2A1, 2A2, 2B1 and 2B2. The first, third, sixth and eighth of these are wins and the other four are losses.

Of course, if the players are allowed to confer, they will simply decide on a code. One reasonable code for the messages sent from the Sender to the Receiver is "A means the state of nature is 1 and B means the state of nature is 2." There is another equally reasonable code, namely "B means the state of nature is 1 and A means the state of nature is 2." Agreeing on either of these beforehand will yield 100 percent efficiency. Even if the players are not allowed to confer, there are good protocols for minimizing the number of times one fails to get a win. One example, which seems likely to occur in real life, would be for the Sender to choose one of the two languages arbitrarily and never deviate, and for the Receiver eventually to conform to it. This requires differentiating the roles in advance

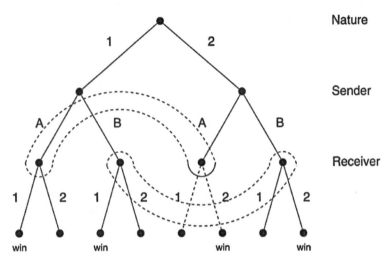

Figure 14.1 The game tree—dashed lines show information sets for the Receiver

but not breaking the symmetry. Such a protocol might be usable, for example, by nodes in a computer network, if the network is bipartite (there are two types of nodes and communication only takes place between nodes of different types). A more general and symmetric protocol might begin with both players playing arbitrarily; subsequent plays could be chosen by copying if the last play in the same situation was successful while choosing randomly otherwise. Here, the definition of "in the same situation" is context dependent but the protocol seems otherwise fairly general.

1.3 An urn scheme for playing the game

The Sender's information set is naturally indexed by Nature's plays: {1, 2}. Correspondingly, the Sender has two urns, call them Urn 1 and Urn 2. Each of these has two colors of ball, call them color A and color B. The contents are initialized to one ball of each color in each urn. Each time the Sender plays, if Nature has played j, then the Sender picks the signal by choosing a ball at random from urn j.

The Receiver plays at four different nodes but, like the Sender, has information partition of size two, indexed by the Sender's plays: {A, B}. Correspondingly, the Receiver has two urns, called urn A and urn B. Each initially contains one ball of color 1 and one ball of color 2.

When the Receiver sees signal x, she chooses an action by picking at random from urn x. Once both players have played, it is revealed whether they won. If they lose, the contents of all urns remain the same, but if they win, the plays are reinforced: each player adds another copy of the ball they chose to the urn it came from. For example, if on the first play the state of nature is 2, the Sender signals A, and the Receiver plays 2 (which, under our model will happen 1/4 of the time that nature plays 2 initially), then a ball of color A is added to the Sender's urn 2 and a ball of color 2 is added to the Receiver's urn A.

We analyze the resulting random sequence of plays under the assumption that Nature chooses states according to independent fair coin flips. When viewed in the classical framework for signaling games, this game has multiple equilibria. In particular, there are two pareto-optimal Nash equilbria corresponding to play according to the two possible languages, and a family of "babbling" equilibria where the Sender ignores the state of nature, choosing signals according to independent coin flips and the Receiver ignores the signal, choosing plays according to her own sequence of independent coin flips. Our goals for this analysis are modest: we show that the urn model protocol converges to one of the two optimal languages in a sense to be made precise in the next section. Note that it is not *a priori* necessary that the urn model produce any Nash equilibrium at all, though we will see in the next section that the model must converge to an appropriately defined set of dynamic equilibria.

1.4 *Formal construction of the model and statement of results*

Let $(\Omega, \mathscr{F}, \mathbb{P})$ be a sufficiently rich source of randomness; for specificity, we take it to be a probability space on which there are defined random variables $\{U_{n,j} : j \in \{1, 2, 3\}, n \geq 1\}$, that are independent and uniform on the unit interval. Let $\mathscr{F}_n = \sigma(U_{k,j} : k < n, j \leq 3)$ be the σ-field of information up to time n. By induction on n, we may simultaneously define random variables corresponding to the contents of the urn at time n and the plays chosen at time n. The variable $V(n, i, x)$ is interpreted as the number of balls of color x in urn i at time n. The induction begins with the initialization $V(n, i, x) = 1$ for $i = 1, 2$ and $x = A, B$ (this populates the two urns used by the Sender) and for $i = A, B$ and $x = 1, 2$ (populating the two urns used by the Receiver).

Given the eight values of $V(n, i, x)$, the plays N_n, S_n and R_n are constructed as follows. Let $N_n = 1$ if $U_{n,1} < 1/2$ and $N_n = 2$ otherwise. The interpretation of N_n is the play chosen by Nature at step n, which is always equally likely to be 1 or 2 independent of the past. Let $S_n = A$ if

$$U_{n,2} < \frac{V(n, N_n, A)}{V(n, N_n, A) + V(n, N_n, B)}$$

and $S_n = B$ otherwise. Thus, conditional on the past, the probability of the Sender choosing signal A is equal to the proportion of balls of color A in the urn N_n, that is, in the urn corresponding to the state that Nature has just played. Similarly, let $R_n = 1$ if

$$U_{n,3} < \frac{V(n, S_n, 1)}{V(n, S_n, 1) + V(n, S_n, 2)}$$

and $R_n = 2$ otherwise, so that the probabilities for the Receiver's plays are proportional to the contents of the urn with the label of the signal at time n.

To complete the induction, update the contents of the urns by defining

$$V(n + 1, i, x) = V(n, i, x) + 1$$

if $N_n = i$ and $S_n = x$ and $R_n = i$ (updating the Sender's urns) or if $S_n = i$ and $R_n = x$ and $N_n = x$ (updating the Receiver's urns), and let $V(n + 1, i, x) = V(n, i, x)$ otherwise.

Our main result may now be stated. Denote the number of wins up through the nth play by $\text{win}_n := \sum_{k=1}^{n} \delta(N_n, R_n)$ where δ is the usual delta function, namely 1 if the arguments are equal and zero otherwise.

THEOREM 1.1. *With probability 1, $\text{win}_n/n \to 1$ as $n \to \infty$. Furthermore, this occurs in one of two specific ways. With probability 1/2, as $n \to \infty$, $V(n, 1, B)/V(n, 1, A)$, $V(n, 2, A)/V(n, 2, B)$, $V(n, A, 2)/V(n, A, 1)$, and $V(n, B, 1)/V(n, B, 2)$ all go to zero, while with probability 1/2, the reciprocals of these all go to zero.*

REMARK 1.2. If arbitrary initial conditions are permitted, that is, if $\{V(n, i, x)\}$ are allowed to be any real vector with strictly positive coordinates, then the same conclusions hold with some probability other than 1/2, measurable in $\mathscr{F}_1$.

Before proceeding to the proof, we make one observation which, though it seems small, reduces the dimension of the problem and simplifies notation considerably. That is, we observe that for each n,

$$V(n+1,1,A) = V(n,1,A) + 1 \Leftrightarrow V(n+1,A,1) = V(n,A,1) + 1$$
$$V(n+1,1,B) = V(n,1,B) + 1 \Leftrightarrow V(n+1,B,1) = V(n,B,1) + 1$$
$$V(n+1,2,A) = V(n,2,A) + 1 \Leftrightarrow V(n+1,A,2) = V(n,A,2) + 1$$
$$V(n+1,2,B) = V(n,2,B) + 1 \Leftrightarrow V(n+1,B,2) = V(n,B,2) + 1.$$

We may therefore keep track of the entire process by keeping track of the four quantities $\{V(n, i, x) : i = 1, 2; x = A, B\}$ instead of all eight quantities. Denoting

$$\mathbf{V}_n := (V(n,1,A), V(n,1,B), V(n,2,A), V(n,2,B))$$

represents the process as a Markov chain $\{V_n\}$. Various formulae will appear more canonical if we refer to the coordinates of $\mathbf{V}_n$ in order as $1A$, $1B$, $2A$, $2B$ instead of $1, 2, 3, 4$, e.g., $(V_n)_{1A} = V(n, 1, A)$ and so forth. If the initial conditions are altered as in Remark 1.2 so that $V(1, i, x) \neq V(1, x, i)$ for some (i, x), then instead of symmetry $V(n, i, x) = V(n, x, i)$, we have that $V(n, i, x) - V(n, x, i)$ is independent of n; the arguments are messier in this case, but the same conclusions hold.

Let $\bowtie$ denote the set $\{1A, 1B, 2A, 2B\}$ and let $T_n := \sum_{j \in \bowtie} V_j$ be the total number of balls in the Sender's urns. Let

$$\mathbf{X}_n := \left(\frac{V(n,1,A)}{T_n}, \frac{V(n,1,B)}{T_n}, \frac{V(n,2,A)}{T_n}, \frac{V(n,2,B)}{T_n} \right) \qquad (1.1)$$

be the normalized proportion vector. The vector $\mathbf{X}_n$ is an element of the interior of the 3-simplex

$$\Delta := \{(x_{1A}, x_{1B}, x_{2A}, x_{2B}) \in \mathbb{R}^{\bowtie} : x_{1A}, x_{1B}, x_{2A}, x_{2B} \geq 0, \sum_{j \in \bowtie} x_j = 1\}.$$

Let us write $X_{n,1,A}$ instead of $(X_n)_{1A}$ and so forth. Let $\psi_n = V_{n+1} = V_n$ be the standard basis vector corresponding to the reinforcement due to the play at time n if there was a win, and the zero vector otherwise. Thus $|\psi_n| = \text{win}_{n+1} - \text{win}_n$, where we use the L^1-norm on $\mathbb{R}^{\bowtie}$ here and throughout.

In this notation, Theorem 1.1 is a consequence of the reformulation

$$\mathbf{X}_n \to \left(\frac{1}{2}, 0, 0, \frac{1}{2}\right) \text{ or } \left(0, \frac{1}{2}, \frac{1}{2}, 0\right). \tag{1.2}$$

That these happen with equal probability when $\mathbf{V}_1 = (1, 1, 1, 1)$ follows from symmetry.

2. Relation to stochastic approximation and an ODE

A common version of the *stochastic approximation* process is one that satisfies

$$\mathbf{X}_{n+1} - \mathbf{X}_n = \gamma_n(F(\mathbf{X}_n) + \xi_n), \tag{2.1}$$

where $\{\gamma_n\}$ are constants such that $\sum_n \gamma_n = \infty$ and $\sum_n \gamma_n^2 < \infty$, and where ξ_n are bounded and $\mathbb{E}(\xi_n \mid F_n) = 0$. Sometimes an extra, possibly random, remainder term R_n is added to $F(\mathbf{X}_n) + \xi_n$, with the condition that $\sum_n \mid R_n \mid < \infty$ almost surely. There is no precise definition for an urn model, but the normalized content vector in an urn model is typically a stochastic approximation processes with $\gamma_n = 1/n$. One sees this by computing $\mathbb{E}(\mathbf{X}_{n+1} - \mathbf{X}_n \mid \mathscr{F}_n)$ and seeing that when scaled by $1/n$ it converges to a vector function F.

To analyze the particular chain $\{\mathbf{V}_n\}$, or equivalently the time-inhomogeneous chain $\{\mathbf{X}_n\}$, begin by writing down the transition probabilities.

$$\mathbb{P}(\psi_n = (1,0,0,0)) = \mathbb{P}(1A1) = \frac{1}{2}\frac{X_{n,1,A}}{X_{n,1,A} + X_{n,1,B}}\frac{X_{n,1,A}}{X_{n,1,A} + X_{n,2,A}};$$

$$\mathbb{P}(\psi_n = (0,1,0,0)) = \mathbb{P}(1B1) = \frac{1}{2}\frac{X_{n,1,B}}{X_{n,1,B} + X_{n,1,A}}\frac{X_{n,1,B}}{X_{n,1,B} + X_{n,2,B}};$$

$$\mathbb{P}(\psi_n = (0,0,1,0)) = \mathbb{P}(2A2) = \frac{1}{2}\frac{X_{n,2,A}}{X_{n,2,A} + X_{n,2,B}}\frac{X_{n,2,A}}{X_{n,2,A} + X_{n,1,A}}; \tag{2.2}$$

$$\mathbb{P}(\psi_n = (0,0,0,1)) = \mathbb{P}(2B2) = \frac{1}{2}\frac{X_{n,2,B}}{X_{n,2,B} + X_{n,2,A}}\frac{X_{n,2,B}}{X_{n,2,B} + X_{n,1,B}};$$

$$\mathbb{P}(\psi_n = (0,0,0,0)) = \mathbb{P}(1*2, 2*1) = 1 - \mathbb{P}(\mid \psi_n \mid = 1),$$

where $*$ denotes a symbol that can be either A or B. Since ψ_n denotes $\mathbf{V}_{n+1} - \mathbf{V}_n$, we have

$$\mathbf{X}_{n+1} - \mathbf{X}_n = \frac{V_{n+1}}{1 + T_n} - \frac{V_n}{1 + T_n} + \frac{V_n}{1 + T_n} - \frac{V_n}{T_n} = \frac{1}{1 + T_n}(\psi_n - \mathbf{X}_n) \quad (2.3)$$

if $|\underline{\psi}_n| = 1$, and $\mathbf{X}_{n+1} - \mathbf{X}_n = \mathbf{0}$ otherwise.

Taking expectations gives

$$\mathbb{E}(\mathbf{X}_{n+1} - \mathbf{X}_n \mid \mathscr{F}_n) = \frac{1}{1 + T_n} F(\mathbf{X}_n), \quad (2.4)$$

where $F(\mathbf{x}) := \mathbb{E}(|\psi_n|(\psi_n - \mathbf{X}_n) \mid \mathbf{X}_n = \mathbf{x})$ is a function from Δ to the tangent space $T\Delta := \{\mathbf{x} \in R^{\bowtie} : \sum_{j \in \bowtie} x_j = 0\}$ given by the formula (written as a column vector so as to fit better)

$$\frac{1}{2}\begin{bmatrix} \frac{(1-x_{1A})x_{1A}^2}{(x_{1A}+x_{1B})(x_{1A}+x_{2A})} - \frac{x_{1A}x_{1B}^2}{(x_{1A}+x_{1B})(x_{1B}+x_{2B})} - \frac{x_{1A}x_{2A}^2}{(x_{2A}+x_{2B})(x_{2A}+x_{1A})} - \frac{x_{1A}x_{2B}^2}{(x_{2B}+x_{2A})(x_{2B}+x_{1B})} \\ -\frac{x_{1B}x_{1A}^2}{(x_{1A}+x_{1B})(x_{1A}+x_{2A})} + \frac{(1-x_{1B})x_{1B}^2}{(x_{1B}+x_{1A})(x_{1B}+x_{2B})} - \frac{x_{1B}x_{2A}^2}{(x_{2A}+x_{2B})(x_{2A}+x_{1A})} - \frac{x_{1B}x_{2B}^2}{(x_{2B}+x_{2A})(x_{2B}+x_{1B})} \\ -\frac{x_{2A}x_{1A}^2}{(x_{1A}+x_{1B})(x_{1A}+x_{2A})} - \frac{x_{2A}x_{1B}^2}{(x_{1B}+x_{1A})(x_{1B}+x_{2B})} + \frac{(1-x_{2A})x_{2A}^2}{(x_{2A}+x_{2B})(x_{2A}+x_{1A})} - \frac{x_{2A}x_{2B}^2}{(x_{2B}+x_{2A})(x_{2B}+x_{1B})} \\ -\frac{x_{2B}x_{1A}^2}{(x_{1A}+x_{1B})(x_{1A}+x_{2A})} - \frac{x_{2B}x_{1B}^2}{(x_{1B}+x_{1A})(x_{1B}+x_{2B})} - \frac{x_{2B}x_{2A}^2}{(x_{2A}+x_{2B})(x_{2A}+x_{1A})} + \frac{(1-x_{2B})x_{2B}^2}{(x_{2B}+x_{2A})(x_{2B}+x_{1B})} \end{bmatrix}.$$

Letting $\xi_n = (1 + T_n)(\mathbf{X}_{n+1} - \mathbf{X}_n - F(\mathbf{X}_n))$ be the noise term, we see that (2.4) is variant of (2.1) with non-deterministic γ_n.

For processes obeying (2.1) or (2.4), the heuristic is that the trajectories of the process should approximate trajectories of the corresponding differential equation $\mathbf{X}' = F(\mathbf{X})$. Let $Z(F)$ denote the set of zeros of the vector field F. The heuristic says that if there are no cycles in the vector field F, then the process should converge to the set $Z(F)$. A sufficient condition for the nonexistence of cycles is that there be a *Liapunov function*, namely a function L such that $\nabla L \cdot F \geq 0$ with equality only where F vanishes. When $Z(F)$ is large enough to contain a curve, there is a question unsettled by the heuristic, as to whether the process can continue to move around in $Z(F)$. There is, however, a nonconvergence heuristic saying that the process should not converge to an unstable equilibrium.

PROPOSITION 2.1 (*Zero Set of F*). *Let Q be the polynomial $x_{1A}x_{2B} - x_{1B}x_{2A}$. The zero set $Z(F)$ of F on Δ consists of the zero set $Z(Q) := \{Q = 0\}$ together with the two points $(\frac{1}{2}, 0, 0, \frac{1}{2})$ and $(0, \frac{1}{2}, \frac{1}{2}, 0)$. $Z(F)$ is shown in* Figure 14.2.

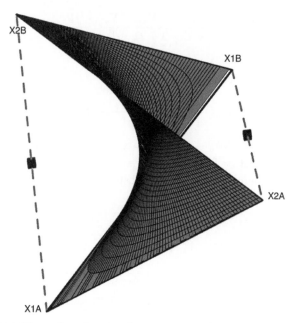

Figure 14.2 The surface $Q = 0$ in barycentric coordinates, and the two other zeros of F

Proof. It is routine to check that F vanishes on the surface and the two points. It suffices, therefore, to check that these are the only solutions to $F = 0$ on Δ. Let Z' be the subset of the simplex where $x_{1A}x_{1B}x_{2A}$ vanishes. In other words, Z' is a union of three of the four faces of Δ. We claim that $Z(F)$ is contained in the set $Z(Q) \cup Z'$.

First, clearing denominators, we let $P_1, \ldots, P_4$ denote the four polynomials obtained by multiplying the components of F by $(x_{1A} + x_{1B})(x_{1A} + x_{2A})(x_{1B} + x_{2B})(x_{2A} + x_{2B})$. Let $P_5 := 1 - x_{1A} - x_{1B} - x_{2A} - x_{2B}$. We will check that $g := Qx_{1A}x_{1B}x_{2A}$ is contained in the *ideal* generated by $P_1, \ldots, P_5$. This is defined as the set of $\sum_{i=1}^{5} q_i P_i$ as q_i range over polynomials, and we denote it by $\mathfrak{J}$. Assuming this for the moment, let us see how the claim is proved. On $Z(F)$, we know that P_5 vanishes because $Z(F) \in \Delta$ and $P_1, \ldots, P_4$ vanishes because F vanishes. Hence every polynomial in $\mathfrak{J}$ vanishes, and in particular g vanishes. The set where g vanishes contains $Z(Q) \cup Z'$, which establishes the claim.

Checking that $g \in \mathfrak{J}$ is easy with the aid of a computer algebra system. For example, in Maple 11 with the Groebner package loaded, the command

$$B := \text{Basis}([P_1, P_2, P_3, P_4, P_5], \text{tdeg}(x_{1A}, x_{1B}, x_{2A}, x_{2B}));$$

produces a **Gröbner basis** for $\mathfrak{J}$ (with respect to the term order tdeg(x_{1A}, x_{1B}, x_{2A}, x_{2B})), this being a canonical representation of $\mathfrak{J}$ for which an algorithm exists to test membership. Specifically, given a polynomial g and a Gröbner basis B, the command to produce a *remainder r* for which $g - r \in B$ and r is small (with respect to the same term order) is

$$\text{Normal Form}(g, B, \text{tdeg}(x_{1A}, x_{1B}, x_{2A}, x_{2B}));$$

When we try this, we find that $r = 0$, implying that $g \in \mathfrak{J}$ and verifying the claim.

Finally, having seen that $Z(F) \subseteq Z(Q) \cup Z'$, identical arguments show that $Z(F) \subseteq Z(Q) \cup Z''$ where Z'' is the zero set in Δ of the product of any three of the four variables x_{1A}, x_{1B}, x_{2A}, x_{2B}. Taking the intersection over the four possible sets Z'' shows that $Z(F) \subseteq Z(Q) \cup Z_*$ where Z_* is the intersection of the zero sets in Δ of the four monomials $x_{1A}x_{1B}x_{2A}$, $x_{1A}x_{1B}x_{2B}$, $x_{1A}x_{2A}x_{2B}$, and $x_{1B}x_{2A}x_{2B}$. In other words, Z_* is the 1-skeleton of Δ (the 1-skeleton being the union of all one-dimensional edges). The set $Z(Q)$ already contains four of the six edges in the 1-skeleton. Checking the edge $(a, 0, 0, 1-a)$ produces exactly one solution to $F = 0$ in the interior of the edge, namely $(\frac{1}{2}, 0, 0, \frac{1}{2})$. Checking the edge $(0, a, 1 - a, 0)$ produces the point $(0, \frac{1}{2}, \frac{1}{2}, 0)$. This finishes the proof of the proposition.

We now check that $Z(Q)$ is a geometrically unstable set for the vector field F.

PROPOSITION 2.2 (*Instability of $Z(Q)$*).

$$\text{sgn}(\nabla Q \cdot F) = \text{sgn}(Q)$$

at all points of Δ, except $(\frac{1}{2}, 0, 0, \frac{1}{2})$ and $(0, \frac{1}{2}, \frac{1}{2}, 0)$.

Proof. The previous proposition shows that F vanishes when Q vanishes, so the conclusion is true when $Q = 0$. By symmetry it suffices to prove that $Q > 0$ implies $\nabla Q \cdot F > 0$ on Δ.

Let $\mathbf{x} = (x_{1A}, x_{1B}, x_{2A}, x_{2B})$ be any point of Δ with $Q(\mathbf{x}) > 0$ and with at most one vanishing coordinate. Then the following relations hold:

$$\frac{x_{1A}}{x_{1A} + x_{1B}} > \frac{x_{2A}}{x_{2A} + x_{2B}};$$

$$\frac{x_{1A}}{x_{1A} + x_{2A}} > \frac{x_{1B}}{x_{2B} + x_{1B}};$$

$$\frac{x_{2B}}{x_{2B} + x_{2A}} > \frac{x_{1B}}{x_{1B} + x_{1A}}; \tag{2.5}$$

$$\frac{x_{2B}}{x_{2B} + x_{1B}} > \frac{x_{2A}}{x_{2A} + x_{1A}}.$$

We may write

$$2\nabla Q \cdot F = x_{1A}x_{2B}H - x_{1B}x_{2A}\tilde{H}, \tag{2.6}$$

where

$$H(\mathbf{x}) = \frac{x_{1A}}{(x_{1A} + x_{1B})(x_{1A} + x_{2A})} + \frac{x_{2B}}{(x_{2B} + x_{2A})(x_{2B} + x_{1B})} - 4\psi(\mathbf{x})$$

$$\tilde{H}(\mathbf{x}) = \frac{x_{2A}}{(x_{2A} + x_{2B})(x_{2A} + x_{1A})} + \frac{x_{1B}}{(x_{1B} + x_{1A})(x_{1B} + x_{2B})} - 4\psi(\mathbf{x}),$$

where $\psi(\mathbf{x}) := \mathbb{P}(|\underline{\psi}_n| = 1 \mid \mathbf{X}_n = \mathbf{x})$.

By the inequalities Eq. (2.5),

$$4\psi(\mathbf{x}) = \frac{2x_{1A}^2}{(x_{1A} + x_{1B})(x_{1A} + x_{2A})} + \frac{2x_{1B}^2}{(x_{1B} + x_{1A})(x_{1B} + x_{2B})}$$

$$+ \frac{2x_{2A}^2}{(x_{2A} + x_{2B})(x_{2A} + x_{1A})} + \frac{2x_{2B}^2}{(x_{2B} + x_{2A})(x_{2B} + x_{1B})}$$

$$< \frac{2x_{1A}^2}{(x_{1A} + x_{1B})(x_{1A} + x_{2A})} + \frac{x_{1A}(x_{1B} + x_{2A})}{(x_{1A} + x_{1B})(x_{1A} + x_{2A})} \tag{2.7}$$

$$+ \frac{x_{2B}(x_{2A} + x_{1B})}{(x_{2B} + x_{2A})(x_{2B} + x_{1B})} + \frac{2x_{2B}^2}{(x_{2B} + x_{2A})(x_{2B} + x_{1B})}$$

$$= \frac{x_{1A}}{x_{1A} + x_{1B}} + \frac{x_{1A}}{x_{1A} + x_{2A}} + \frac{x_{2B}}{x_{2B} + x_{2A}} + \frac{x_{2B}}{x_{2B} + x_{1B}}.$$

Denote the common denominator

$$D := (x_{1A} + x_{1B})(x_{1A} + x_{2A})(x_{2B} + x_{2A})(x_{2B} + x_{1B}). \tag{2.8}$$

It follows (using $x_{1A} + x_{1B} + x_{2A} + x_{2B} = 1$ in the second line) that

$$H(\mathbf{x}) > \frac{x_{1A}(1 - (x_{1A} + x_{1B}) - (x_{1A} + x_{2A}))}{(x_{1A} + x_{1B})(x_{1A} + x_{2A})}$$

$$+ \frac{x_{2B}(1 - (x_{2B} + x_{2A}) - (x_{2B} + x_{1B}))}{(x_{2B} + x_{2A})(x_{2B} + x_{1B})}$$

$$= \frac{x_{1A}(x_{2B} - x_{1A})}{(x_{1A} + x_{1B})(x_{1A} + x_{2A})} + \frac{x_{2B}(x_{1A} - x_{2B})}{(x_{2B} + x_{2A})(x_{2B} + x_{1B})}$$

$$= (x_{1A} + x_{2B}) \left[\frac{x_{2B}}{(x_{2B} + x_{2A})(x_{2B} + x_{1B})} - \frac{x_{1A}}{(x_{1A} + x_{1B})(x_{1A} + x_{2A})} \right]$$

$$= \frac{(x_{1A} - x_{2B})^2 Q}{D} > 0.$$

Analogous computations show that $\tilde{H} < 0$. Since at most one of the coordinates vanishes, it follows that the left-hand side of (3.5) is strictly positive.

Finally, if more than one of the coordinates of $\mathbf{x}$ vanishes but $Q \neq 0$, then $\mathbf{x}$ is interior to one of the two line segments $(a, 0, 0, 1 - a)$ or $(0, a, 1 - a, 0)$. Plugging in these parameterizations shows the only interior zeros of $\nabla Q \cdot F$ to be at the midpoints.

3. Probabilistic analysis

Lemma 3.1. *With probability* 1,

$$\frac{1}{2} \leq \liminf \frac{T_n}{n} \leq \limsup \frac{T_n}{n} \leq 1.$$

Proof. The upper is trivial because $T_n \leq n - 1 + T_1$. The lower bound follows from the conditional Borel–Cantelli lemma [Durrett 2004: Theorem I.6] once we show that $\psi(\mathbf{x})$ is always at least 1/2. To prove the lower bound, multiply the expression (2.7) for ψ by D to clear the denominators, and double. The result is easily seen to be $D + Q^2$. Thus

$$\psi - \frac{1}{2} = \frac{Q^2}{2D},$$

which is clearly a nonnegative quantity.   □

With this preliminary result out of the way, the remainder of the proof of Theorem 1.1 may be broken into three pieces, namely Propositions 3.2–3.4. We have seen that $L := Q^2$ is a Liapunov function for the stochastic process $\{X_n\}$; this is implied by Proposition 2.2 and the fact that $\nabla(Q^2)$ is parallel to ∇Q. The minimum value of zero occurs exactly on the surface $Z(Q)$ and the maximum value of $1/16$ occurs at the two other points of $Z(F)$. Let

$$Z_0(Q) := Z(Q) \cap \partial\Lambda = Z(D).$$

PROPOSITION 3.2 (*Liapunov Function Implies Convergence*). *The stochastic process $\{L(X_n)\}$ converges almost surely to 0 or 1/16.*

PROPOSITION 3.3 (*Instability Implies Nonconvergence*). *The probability that $\lim_{n\to\infty} X_n$ exists and is in $Z(Q) \setminus Z_0(Q)$ is zero.*

PROPOSITION 3.4 (*No Convergence to Boundary*). *The limit $\lim_{n\to\infty} X_n$ exists with probability 1. Furthermore, $\mathbb{P}\left(\lim_{n\to\infty} X_n \in Z_0(Q)\right) = 0$.*

These three results together imply Theorem 1.1. The first is an easy result; it is shown via martingale methods that $\{X_n\}$ cannot continue to cross regions where F does not vanish. The second result, fashioned after the non-convergence results of Pemantle (1990: Theorem 1) and generalizations such as Benaïm (1999: Theorem 9.1), follows the argument, by now standard, given in condensed form in Pemantle (2007: Theorem 2.9). The third result is the trickiest, relying on special properties of the process $\{X_n\}$. This is necessary because the nonconvergence method of Pemantle (1990) fails near the boundary of an urn scheme due to diminished variance of the increments; a more general rubric for proving nonconvergence to unstable points in such cases and proving convergence of the process (and not just the Liapunov function) would be desirable.

Proof of Proposition 3.2. Denote $Y_n := L(X_n)$. Decompose $\{Y_n\}$ into a martingale and a predictable process $Y_n = M_n + A_n$ where $A_{n+1} - A_n = \mathbb{E}(Y_{n+1} - Y_n \mid F_n)$. The increments in Y_n are $O(1/T_n) = O(1/n)$ almost surely by Lemma 3.1, so the martingale $\{M_n\}$ is in L^2 and hence almost surely convergent. To evaluate A_n, use the Taylor expansion

$$L(\mathbf{x} + \mathbf{y}) = L(\mathbf{x}) + \mathbf{y} \cdot \nabla L(\mathbf{x}) + R_{\mathbf{x}}(\mathbf{y})$$

with $R_{\mathbf{x}}(\mathbf{y}) = O(|\mathbf{y}|^2)$ uniformly in $\mathbf{x}$. Then

$$
\begin{aligned}
A_{n+1} - A_n &= \mathbb{E}[L(F(\mathbf{X}_{n+1})) - L(F(\mathbf{X}_n)) \mid \mathscr{F}_n] \\
&= \mathbb{E}[\nabla L(\mathbf{X}_n) \cdot (\mathbf{X}_{n+1} - \mathbf{X}_n) + R_{\mathbf{X}_n}(\mathbf{X}_{n+1} - \mathbf{X}_n) \mid \mathscr{F}_n] \\
&= \frac{1}{1 + T_n}(\nabla L \cdot F)(\mathbf{X}_n) + \mathbb{E}[R_{\mathbf{X}_n}(\mathbf{X}_{n+1} - \mathbf{X}_n) \mid \mathscr{F}_n].
\end{aligned}
$$

Since the $R_{\mathbf{X}_n}(\mathbf{X}_{n+1} - \mathbf{X}_n) = O(T_n^{-2}) = O(n^{-2})$ is summable, this gives

$$A_n = \eta(n) + \sum_{k=1}^{n} \frac{1}{1 + T_k}(\nabla L \cdot F)(\mathbf{X}_k)$$

for some almost surely convergent η.

We may now use the usual argument by contradiction: if $\mathbf{X}_n$ is found infinitely often away from the critical values of the Liapunov function, then the drift would cause the Liapunov function to blow up. To set this up, observe first that boundedness of $\{Y_n\}$ and $\{M_n\}$ imply that $\{A_n\}$ is bounded. For any $\epsilon \in (0, 1/32)$, let Δ_ϵ denote $L^{-1}[\epsilon, 1/16 - \epsilon]$. On Δ_ϵ, the function $\nabla L \cdot F$, which is always nonnegative, is bounded below by some constant c_ϵ. Let δ be the distance from Δ_ϵ to the complement of $\Delta_{\epsilon/2}$. Suppose $\mathbf{X}_n, \mathbf{X}_{n+1}, \ldots, \mathbf{X}_{n+k-1} \in \Delta_\epsilon$ and $\mathbf{X}_{n+k} \notin \Delta_{\epsilon/2}$. Then, since $|\psi_n|$ and $|\mathbf{X}_n|$ are at most 1, from (2.3) we see that

$$
\begin{aligned}
\delta &\le \sum_{j=n}^{n+k-1} | \mathbf{X}_{j+1} - \mathbf{X}_j | \\
&\le \sum_{j=n}^{n+k-1} \frac{2}{1 + T_j} \\
&\le \frac{4}{\epsilon}[A_{n+k} - A_n - (\eta(n+k) - \eta(n))].
\end{aligned}
$$

Thus, if $\mathbf{X}_n \in \Delta_\epsilon$ infinitely often, it follows that $\{A_n\}$ increases without bound. By contradiction, for each ϵ, $\{\mathbf{X}_n\}$ eventually remains outside of Δ_ϵ, which proves the proposition.

Proof of Proposition 3.3. The idea of this proof appeared first in Pemantle (1988: 103), but the hypotheses there, as well as those of Pemantle (1990: Theorem 1) and Benaïm (1999: Theorem 9.1) require deterministic step sizes $\{\gamma_n\}$ and analyses of isolated unstable fixed points or entire unstable orbits. We therefore take some care here to document what is minimally required of the process $\{X_n\}$ and its Liapunov function Q.

For any process $\{Y_n\}$ we let ΔY_n denote $Y_{n+1} - Y_n$. Let $N \subseteq \mathbb{R}^d$ be any closed set, let $\{X_n : n \geq 0\}$ be a process adapted to a filtration $\{\mathscr{F}_n\}$ and let $\sigma := \inf\{k : X_k \notin \mathcal{N}\}$ be the time the process exits $\mathcal{N}$. Let $\mathbb{P}_n$ and $\mathbb{E}_n$ denote conditional probability and expectation with respect to F_n. We will impose several hypotheses, (3.1)–(3.3), on $\{X_n\}$ and then check that the process $\{X_n\}$ defined in (1.1) satisfies these conditions. We require

$$\mathbb{E}_n \mid \Delta X_n \mid^2 \leq c_1 n^{-2} \tag{3.1}$$

for some constant $c_1 > 0$, which also implies $\mathbb{E}_n \mid \Delta X_n \mid \leq \sqrt{c_1} n^{-1}$. Let Q be a twice differentiable real function on a neighborhood $\mathcal{N}'$ of $\mathcal{N}$. We require that

$$\operatorname{sgn}(Q(X_n))[\nabla Q(X_n) \cdot \mathbb{E}_n \Delta X_n] \geq -c_2 n^{-2} \tag{3.2}$$

whenever $X_n \in \mathcal{N}'$. Let c_3 be an upper bound for the determinant of the matrix of second partial derivatives of Q on $\mathcal{N}'$. We require a lower bound on the incremental variance of the process $\{Q(X_n)\}$:

$$\mathbb{E}_n(\Delta Q(X_n))^2 \geq c_4 n^{-2} \tag{3.3}$$

when $n < \sigma$. The relation between these assumptions and the process $\{X_n\}$ defined in (1.1) is as follows.   □

LEMMA 3.5. *Suppose there is a function $F : \mathcal{N} \to T\Delta$ and there are nonnegative quantities $\gamma_n \in \mathscr{F}_n$ and $c' > 0$ such that*

$$\mid \mathbb{E}_n \Delta X_n - \gamma_n F(X_n) \mid \leq c' n^{-2}; \tag{3.4}$$

$$\operatorname{sgn}(\nabla Q \cdot F) = \operatorname{sgn}(Q). \tag{3.5}$$

Then (3.2) is satisfied. When N is disjoint from $\partial \Delta$, it follows that the particular process $\{X_n\}$ defined in (1.1) satisfies (3.1) and (3.3) as well as (3.2).

Proof. Let $R := \mathbb{E}_n \Delta X_n - \gamma_n F(X_n)$. Then

$$\nabla Q(X_n) \cdot \mathbb{E}_n \Delta X_n = \nabla Q(X_n) \cdot [\gamma_n F(X_n) + R]$$
$$\geq 0- \mid \nabla Q(X_n) \mid c' n^{-2}.$$

and (3.2) follows by picking $c_2 \geq c'$ $\sup_{x \in \mathcal{N}} |\nabla Q(x)|$. The process $\{X_n\}$ of (1.1) satisfies (3.1) because $|\Delta X_n|$ is bounded from above by n^{-1}. Finally, to see (3.3), note that $|\nabla Q| \geq \epsilon > 0$ on any closed set disjoint from $\partial \Delta$, and also that on such a set $\mathbb{P}(\psi_n = e_j)$ is bounded from below for any elementary basis vector e_j; the lower bound on the second moment of ΔQ (X_n) follows. $\square$

Proposition 3.3 now follows from a more general result:

PROPOSITION 3.6. *Let $\{X_n\}$, Q, $\mathcal{N} \subseteq \mathcal{N}'$ and the exit time σ from $\mathcal{N}$ be defined as in the proof of Proposition 3.3 and satisfy (3.1)–(3.3), with constants c_1, c_2, and c_4 appearing there and the bound c_3 on the Hessian determinant of Q on $\mathcal{N}'$. Assume further that there is an N_0 such that for $n \geq N_0$, $X_n \in \mathcal{N} \Rightarrow X_{n+1} \in \mathcal{N}'$. Then*

$$\mathbb{P}[\sigma = \infty \text{ and } Q(X_n) \to 0] = 0.$$

REMARK. Proposition 3.3 follows by applying this to a countable cover of $Z(Q) \setminus Z_0(Q)$ by compact sets.

Proof. The structure of the proof mimics the nonconvergence proofs of Pemantle (1988, 1990) and Benaïm (1999). We show that the incremental quadratic variation of the process $\{Q(X_n)\}$ is of order at least n^{-2}; this is (3.7). Then we show that conditional on any past at time n, the probability is bounded away from zero that the process $\{Q(X_n)\}$ wanders away from zero by at least a constant multiple of $n^{-1/2}$ (this is Lemma 3.7) and that the subsequent probability of never returning much nearer to zero is also bounded from below (this is Lemma 3.8).

To begin in earnest, we fix $\epsilon > 0$ and $N \geq N_0$ also satisfying

$$N \geq \frac{16(c_2 + c_1 c_3)^2}{c_4^2}. \tag{3.6}$$

Let $\tau := \inf\{k \geq N : |Q(X_k)| \geq \epsilon k^{-1/2}\}$. Suppose that $N \leq n \leq \sigma \wedge \tau$. From the Taylor estimate

$$| Q(x + y) - Q(x) - \nabla Q(x) \cdot y | \geq C | y |^2,$$

where C is an upper bound on the Hessian determinant for Q on the ball of radius $|y|$ about x, we see that

$$\mathbb{E}_n \Delta (Q(\mathbf{X}_n)^2) = \mathbb{E}_n 2Q(\mathbf{X}_n) \Delta Q(\mathbf{X}_n) + \mathbb{E}_n (\Delta Q(\mathbf{X}_n))^2$$
$$\geq 2Q(\mathbf{X}_n) \nabla Q(\mathbf{X}_n) \cdot \mathbb{E}_n \Delta \mathbf{X}_n - 2c_3 Q(\mathbf{X}_n) \mathbb{E}_n |\Delta \mathbf{X}_n|^2 + \mathbb{E}_n |\Delta Q(\mathbf{X}_n)|^2$$
$$\geq [-2Q(\mathbf{X}_n)(c_2 + c_3 c_1) + c_4] n^{-2}.$$

By (3.6), we have $n^{-1/2} \leq c_4/(4(c_2 + c_1 c_3))$. Hence

$$\mathbb{E}_n \Delta (Q(\mathbf{X}_n)^2) \geq \frac{c_4}{2} n^{-2}. \tag{3.7}$$

Lemma 3.7. *If ε is taken to equal $c_4/2$ in the definition of τ, then* $\mathbb{P}_n(\tau \wedge \sigma < \infty) \geq 1/2$.

Proof. For any $m \geq n$ it is clear that $|Q(\mathbf{X}_{m \wedge \tau \wedge \sigma})| \leq \varepsilon n^{-1/2}$. Thus,

$$\varepsilon n^{-1} \geq \mathbb{E}_n Q(\mathbf{X}_{m \wedge \tau \wedge \sigma})^2$$
$$\geq \mathbb{E}_n \left[Q(\mathbf{X}_{m \wedge \tau \wedge \sigma})^2 - Q(\mathbf{X}_n)^2 \right]$$
$$= \sum_{k=n}^{m-1} \mathbb{E}_n \Delta (Q(\mathbf{X}_k)^2) \mathbf{1}_{k < \tau \wedge \sigma}$$
$$\geq \sum_{k=n}^{m=1} c_4 n^{-2} P_n(\sigma \wedge \tau > k)$$
$$\geq \frac{c_4}{2} (n^{-1} - m^{-1}) \mathbb{P}_n(\sigma \wedge \tau = \infty).$$

Letting $m \to \infty$ we conclude that $\epsilon \leq c_4/2$ implies $\mathbb{P}(\tau \wedge \sigma = \infty) \leq 1/2$.

LEMMA 3.8. *There is an N_0 and a $c_5 > 0$ such that for all $n \geq N_0$,*

$$\mathbb{P}_n \left(\sigma < \infty \text{ or for all } m \geq n, |Q(\mathbf{X}_m)| \geq \frac{c_4}{5} n^{-1/2} \right) \geq c_5$$

whenever $|Q(\mathbf{X}_n)| \geq (c_4/2) n^{-1/2}$.

Let us now see that Lemmas 3.7 and 3.8 prove Proposition 3.6. Let $\mathcal{N}$ be any closed ball in the interior of Δ and let $\mathcal{N}'$ be any convex neighborhood of $\mathcal{N}$ whose closure is still in the interior of Δ. For $n \geq N_0$, we have

$$\mathbb{P}_n[\sigma = \infty \text{ and } Q(\mathbf{X}_n) \to 0] \frac{1}{2} + \frac{1}{2}(1 - c_5) < 1.$$

But $\mathbb{P}_n(A) \to \mathbf{1}_A$ almost surely for any event $A \in \sigma(\cup_n \mathscr{F}_n)$. Thus

$$\mathbb{P}_n[\sigma = \infty \text{ and } Q(\mathbf{X}_n) \to 0] \to 1$$

almost surely on the event $\{\sigma = \infty \text{ and } Q(\mathbf{X}_n) \to 0\}$, and it follows that the probability of this event is zero. It remains to prove Lemma 3.8.

Let $\phi(x) := \phi_\lambda(x) := x + \lambda x^2$ and let $\tilde{Q}(x) := \phi(Q(x))$. First, we establish that there is a $\lambda > 0$ such that $\tilde{Q}(X_n)$ is a submartingale when $Q \geq 0$ and $n \geq N_0$.

$$\mathbb{E}_n \Delta \tilde{Q}(X_n) = \mathbb{E}_n \Delta Q(X_n) + \lambda \mathbb{E}_n \Delta (Q(X_n)^2)$$
$$\geq \nabla Q(X_n) \cdot \mathbb{E}_n \Delta X_n - c_3 \mathbb{E}_n |\Delta X_n|^2 + \lambda \frac{c_4}{2} n^{-2}.$$

Choosing $\lambda \geq (2/c_4)(c_2 + c_1 c_3)$ then yields a submartingale when $Q(X_n) \geq 0$.

Next, let $M_n + A_n$ denote the Doob decomposition, of $\{\tilde{Q}(X_n)\}$; in other words, $\{M_n\}$ is a martingale and A_n is predictable and increasing. An upper bound on $|\phi_\lambda'(Q(x))|$ is $c_7 := 1 + 2\lambda \sup |Q| = 1 + 2\lambda$. From the definition of Q, we see that $|\nabla Q| \leq 1$. It follows from these two facts that

$$\frac{|\tilde{Q}(x+y) - \tilde{Q}(x)|}{|y|} \leq 1 + 2\lambda.$$

It is now easy to estimate that

$$\mathbb{E}_n (\Delta M_n)^2 \leq \mathbb{E}_n (\Delta \tilde{Q}(X_n))^2$$
$$\leq \left(\sup \frac{|\tilde{Q}(x+y) - \tilde{Q}(x)|}{|y|} \right) \mathbb{E}_n |\Delta X_n|^2$$
$$\leq c_1 c_7 n^{-2} \sup \frac{d\tilde{Q}}{dQ}.$$

We conclude that there is a constant $c_6 > 0$ such that $\mathbb{E}_n (\Delta M_n)^2 \leq c_6 n^{-2}$ and consequently

$$\mathbb{E}_n (M_{n+m} - M_n)^2 \leq c_6 n^{-1} \tag{3.8}$$

for all $m \geq 0$ on the event $\{Q(X_n) \geq 0\}$.

For any $a, n, V > 0$ and any martingale $\{M_k\}$ satisfying $M_n \geq a$ and $\sup_m \mathbb{E}_n (M_{n+m} - M_n)^2 \leq V$, there holds an inequality

$$\mathbb{P}\left(\inf_m M_{n+m} \leq \frac{a}{2} \right) \leq \frac{4V}{4V + a^2}.$$

To see this, let $\tau = \inf\{k \geq n : M_k \leq a/2\}$ and let $p := \mathbb{P}_n(\tau < \infty)$. Then

$$V \geq p\left(\frac{a}{2}\right)^2 + (1-p)\mathbb{E}_n(M_\infty - M_n \mid \tau = \infty)^2 \geq p\left(\frac{a}{2}\right)^2 + (1-p)\left(\frac{p(a/2)}{1-p}\right)^2$$

which is equivalent to $p \leq 4V/(4V + a^2)$. It follows, with $a = \frac{c_4}{2}n^{-1/2}$ and $V = c_6 n^{-1}$, that

$$\mathbb{P}_n\left(\inf_{k \geq n} M_k \leq \frac{c_4}{4}n^{-1/2}\right) \leq c_5 := \frac{4c_6}{4c_6 + (1/4)c_4^2}.$$

But $M_k \leq \tilde{Q}(\mathbf{X}_k)$ for $k \geq n$, so $Q(\mathbf{X}_k) \leq (c_4/5)n^{-1/2}$ implies $\tilde{Q}(\mathbf{X}_k) \leq (c_4/4)n^{-1/2}$ for $n \geq N_0$, which implies $M_k \leq (c_4/4)n^{-1/2}$. Thus the conclusion of the lemma is established in the positive case, $Q(\mathbf{X}_n) \geq (c_4/2)n^{-1/2}$. An entirely analogous computation shows that $Q(\mathbf{X}_n) - \lambda Q(\mathbf{X}_n)^2$ is a supermartingale when $Q(\mathbf{X}_n) \leq 0$, and the conclusion follows as well in the negative case, that is, the case $Q(\mathbf{X}_n) \geq (c_4/2)n^{-1/2}$. The lemma is established, and along with it, Proposition 3.3.

Proof of Proposition 3.4. The following lemma compares an urn process to a Pólya urn, deducing from the known properties of Pólya's urn that the compared urn satisfies an inequality. The proof is easy and consumes space only in order to spell out certain couplings.

LEMMA 3.9. *Suppose an urn has balls of two colors, white and black. Suppose that the number of balls increases by precisely 1 at each time step, and denote the number of white balls at time n by W_n and the number of black balls by B_n. Let $X_n := W_n/(W_n + B_n)$ denote the fraction of white balls at time n and let F_n denote the σ-field of information up to time n. Suppose further that there is some $0 < p < 1$ such that the fraction of white balls is always attracted toward p:*

$$(\mathbb{P}(X_{n+1} > X_n \mid F_n) - X_n) \cdot (p - X_n) \geq 0.$$

Then the limiting fraction $\lim_{n \to \infty} X_n$ almost surely exists and is strictly between zero and one.

Proof. Let $\tau_N := \inf\{k \geq N : X_k \leq p\}$ be the first time after N that the fraction of white balls drops below p. The process $\{X_{k \wedge \tau_N} : k \geq N\}$ isa bounded supermartingale, hence converges almost surely. Let $\{(W'_k, B'_k) : k \geq N\}$ be a Pólya urn process coupled to $\{(W_k, B_k)\}$ as follows. Let $(W'_N, B'_N) = (W_N, B_N)$. We will verify inductively that $X_k \leq X'_k := W'_k/(W'_k + B'_k)$ for all $k \leq \tau_N$. If $k < \tau_N$ and $W_{k+1} - W_k = 1$ then let $W'_{k+1} = W'_k + 1$. If $k < \tau_N$ and $W_{k+1} = W_k$ then let Y_{k+1} be a Bernoulli random variable independent of everything else with $\mathbb{P}(Y_{k+1} = 0 \mid \mathscr{F}_k) = (1 - X'_k)/(1 - X_k)$, which is nonnegative. Let $W'_{k+1} = W_k + Y_{k+1}$. The construction guarantees that $W'_{k+1} \geq X_{k+1}$, completing the induction, and it is easy to see that $\mathbb{P}(W'_{k+1} > W_k) = X'_k$, so that $\{X'_k : N \leq k \leq \tau_N\}$ is a Pólya urn process.

Complete the definition by letting $\{X'_k\}$ evolve independently as a Pólya urn process once $k \geq \tau_N$. It is well-known that X'_k converges almost surely and that the conditional law of $X'_\infty := lim_{k\to\infty} X_k$ given F_N is a beta distribution, $\beta(W_N, B_N)$. For later use, we remark that beta distributions satisfy the estimate

$$\mathbb{P}(|\beta(xn, (1 - x)n) - x \mid >; \delta) \leq c_1 e^{-c_2 n\delta} \tag{3.9}$$

uniformly for x in a compact sub-interval of $(0, 1)$ Since the beta distribution has no atom at 1, we see that $lim_{k\to\infty} X_k$ is strictly less than 1 on the event $\{\tau_N = \infty\}$. An entirely analogous argument with τ_N replaced by $\sigma_N := \inf\{k \geq N : X_k \geq p\}$ shows that $lim_{k\to\infty} X_k$ is strictly greater than 0 on the event $\{\sigma_N = \infty\}$. Taking the union over N shows that $lim_{k\to\infty} X_k$ exists on the event $\{(X_k - p)(X_{k+1} - p) < 0$ finitely often$\}$ and is strictly between zero and one. The proof of the lemma will therefore be finished once we show that $X_k \to p$ on the event that $X_k - p$ changes sign infinitely often.

Let $G(N, \epsilon)$ denote the event that $X_{N-1} < p < X_N$ and there exists $k \in [N, \tau_N]$ such that $X_k > p + \epsilon$. Let $H(N, \epsilon)$ denote the event that $X_{N-1} > p > X_N$ and there exists $k \in [N, \sigma_N]$ such that $X_k < p + \epsilon$. It suffices to show that for every $\epsilon > 0$, the sums $\sum_{N=1}^\infty \mathbb{P}(G(N, \epsilon))$ and $\sum_{N=1}^\infty \mathbb{P}(H(N, \epsilon))$ are finite; for then by Borel–Cantelli, these occur finitely often, implying $p - \epsilon \leq \lim \inf X_k \leq \lim \sup X_k \leq p + \epsilon$ on the event that $X_k - p$ changes sign infinitely often; since ϵ is arbitrary, this suffices. Recall the Pólya urn coupled to $\{X_k : N \leq k \leq \tau_N\}$. On the event $G(N, \epsilon)$, either $X'_\infty \geq \epsilon/2$ or $X'_\infty - X_\rho \leq -\epsilon/2$ where $\rho \geq k$ is the least $m \geq N$ such that $S'_m \geq \epsilon$. The conditional distribution of $X'_\infty - X_\rho$ given F_ρ is $\beta(W'_\rho, B'_\rho)$. Hence

$$\begin{aligned}
\mathbb{P}(G(N, \epsilon)) \leq {} &\mathbb{E}1_{X_{N-1} < p < X_N} \mathbb{P}\left(\beta(W_N, B_N) \geq \frac{\epsilon}{2}\right) \\
&+ \mathbb{E}1_{\rho < \infty} \mathbb{P}\left(\beta(W'_\rho, B'_\rho) \leq -\frac{\epsilon}{2}\right).
\end{aligned} \tag{3.10}$$

Combining this with the estimate (3.9) establishes summability of $\mathbb{P}(G(N, \epsilon))$. An entirely analogous argument establishes summability of $\mathbb{P}(H(N, \epsilon))$, finishing the proof of the lemma.

Proof of Proposition 3.4. Color the urn process $\{V_n\}$, by coloring balls of types 1A and 1B white and coloring balls of type 2A and 2B black. Let $\tau_k := \inf\{k : T_n = k\}$ denote the times of increase of $\{T_n\}$. We let $W_k := V(\tau_k, 1, A) + V(\tau_k, 2, B)$ denote the number of white balls at time τ_k and $B_k := V(\tau_k, 2, A) + V(\tau_k, 1, B)$ denote the number of black balls. We claim that the urn process $\{(W_k, B_k)\}$ satisfies the hypotheses of Lemma 3.9 with $p = 1/2$. To verify this, let $(x_{1A}, x_{1B}, x_{2A}, x_{2B})$ denote X_{τ_n} and write $\mathbb{P}(X_{n+1} > X_n \mid \mathscr{F}_n) - X_n$ as Num/Den where

$$\text{Num} = \frac{x_{1A}^2}{(x_{1A} + x_{1B})(x_{1A} + x_{2A})} + \frac{x_{1B}^2}{(x_{1A} + x_{1B})(x_{2B} + x_{1B})};$$

$$\text{Den} = \text{Num} + \frac{x_{2A}^2}{(x_{2B} + x_{2A})(x_{1A} + x_{2A})} + \frac{x_{2B}^2}{(x_{2B} + x_{1A})(x_{2B} + x_{1B})}.$$

Simplifying and using $x_{1A} + x_{1B} + x_{2A} + x_{2B} = 1$ shows that

$$\mathbb{P}(X_{n+1} > X_n \mid \mathscr{F}_n) - X_n = -\frac{(x_{1,A} + x_{1,B} - x_{2,A} - x_{2,B})Q^2}{(x_{1,A} + x_{1,B} + x_{2,A} + x_{2,B})(Q^2 + D)},$$

where, as before, D denotes the common denominator (2.8). This is clearly nonpositive when $x_{1A} + x_{1B} \geq x_{2A} + x_{2B}$. This is the same condition as $x_{1A} + x_{1B} \geq 1/2$, so the claim is proved.

Lemma 3.9 now allows us to conclude that $(V(n, 1, A) + V(n, 1, B))/T_n$ converges to a non-zero value. The process $\{V_n : n \geq 0\}$ is invariant under transposing the first and fourth coordinates, as also under transposing the second and third coordinates. We conclude that the four quantities

$$\frac{V(n, 1, A) + V(n, 1, B)}{T_n}, \quad \frac{V(n, 1, A) + V(n, 2, A)}{T_n},$$

$$\frac{V(n, 2, B) + V(n, 2, A)}{T_n}, \quad \frac{V(n, 2, B) + V(n, 1, B)}{T_n} \qquad (3.11)$$

all converge almost surely to nonzero values.

Combining this with Proposition 3.2, we see that there is almost surely a pair of numbers $a, b \in (0, 1)$ such that the limit set of V_n is contained in the set

$$\Xi_{a,b} := \left\{ x := (x_{1A}, x_{1B}, x_{2A}, x_{2B}) \in \Delta : L(x) \in \left\{ 0, \frac{1}{16} \right\} \text{ and} \right.$$

$$\left. x_{1A} + x_{1B} = a \text{ and } x_{1A} + x_{2A} = b \right\}.$$

When $a = b = 1/2$, the set $\Xi_{a,b}$ consists of the three points $(1/2, 0, 0, 1/2)$, $(0, 1/2, 1/2, 0)$ and $(1/4, 1/4, 1/4, 1/4)$. In any other case, the set $\{x_{1A} + x_{1B} = a, x_{1A} + x_{2A} = b\}$ in the simplex Δ is a line segment parallel to $(1, -1, -1, 1)$, and can never intersect $\{Q = 0\}$ in more than one point, hence the set $\Xi_{a,b}$ consists of at most one point. Almost sure convergence of V_n follows.

On $Z_0(Q)$, one of the four quantities $x_{1A} + x_{1B}, x_{1A} + x_{2A}, x_{2B} + x_{2A}, x_{2B} + x_{1B}$ always vanishes; according to (3.11), the law of $\lim_{n \to \infty} V_n$ must therefore give zero probability to the set $Z_0(Q)$, establishing the final conclusion of the proposition, and finishing the proof of Theorem 1.1.

4. Discussion

We have analyzed what we consider to be the simplest nontrivial model of a coordination game. There are a number of natural extensions to the model, all of which raise interesting questions and none of which has been rigorously analyzed. A list of extensions for which we have both simulations and heuristics (via an ODE) but no rigorous analyses is as follows: states not equally probable; number of states, signals, or acts greater than two (the problems differ depending on which of the numbers is greatest); more than two agents interacting in a signaling network. We consider these in turn.

Suppose we have 2 states, 2 acts, and 3 signals. Do we still get efficient signaling? Does one signal fall out of use so that we end up with essentially a two-signal system, or does one signal comes to stand for one state and the other two persist as synonyms for the other state? Heuristics and simulations suggest that synonyms form, with no signal falling out of use. Suppose we have 3 states, 3 acts, and 2 signals. There is now an informational bottleneck and efficient signaling is right only 2/3 of the time. Again efficiency could be achieved in different ways. It appears that one signal is shared between two states, rather than one state being left without a signal. Moving beyond two agents, suppose that there are two senders and one receiver. There are 4 states, but each sender only observes the correct member of a partition. Sender 1 observes the partition {{1, 2}, {3, 4}} and Sender 2 observes the partition {{1, 3}, {2, 4}}. Each sender has 2 signals and the receiver has 4 acts, each paying off in exactly one state, and in that case everyone is reinforced. On the other hand we can have one sender and two receivers. The sender observes one of 4 states, and sends one of 2 signals to each receiver. The receivers each choose one of two acts, and the pair of acts chosen must be right for the state for everyone to be reinforced. We can have a chain, where the sender observes one of 2 states, sends one of 2 signals to an intermediary and the intermediary sends one of 2 signals to the receiver. The receiver must do one of two acts, and if it is right for the state all get reinforced. Simulations suggest that in each of the models described in this paragraph, individuals always learn to signal.

However, even simpler variations may introduce new complexity. With 3 states, 3 signals, and 3 acts, there is a new class of equilibria of partial information transfer, which combines bottlenecks and synonyms. For example, the sender always sends signal 1 in states 1 and 2 and mixes between signals 2 and 3 in state 3. The receiver always does acts 3 when getting signals 2 and 3, and mixes between acts 1 and 2 when getting signal 1. Simulations suggest that reinforcement sometimes converges to such equilibria and sometimes to signaling systems. The slow convergence of such systems to equilibrium behavior casts some doubt on whether these mixed equilibria are in fact possible (cf. Pemantle and Volkov 1999: Theorem 1.2 and the remark following; see also Pemantle and Skyrms 2001).

PROBLEM. Determine whether mixed equilibria are possible in the case of 3 states, 3 signals, and 3 acts.

Finally, if we lift the assumption that states are equiprobable, simulations suggest that even in the 2 state, 2 signal, 2 act case it is possible for reinforcement to converge to a state where the receiver ignores the signal and always chooses the act that is right for the most probable state. In these cases, recovery of almost sure convergence to efficient signaling may require some perturbation of the learning dynamics.

Acknowledgement

Robin Pemantle's research was supported in part by National Science Foundation grant # DMS 0103635.

References

Alexander, J. (2007) *The Structural Evolution of Morality*, Cambridge: Cambridge University Press.

Benaïm, M. (1999) "Dynamics of stochastic approximation algorithms." In *Seminaires de Probabilités XXXIII*, Lecture Notes in Mathematics, vol. 1709, 1–68. Berlin: Springer.

Bonacich, P. and T. Liggett (2003) "Asymptotics of a matrix valued Markov chain arising in sociology." *Stochastic Process Applications* 104: 155–71.

Bradt, R., S. Johnson, and S. Karlin (1956) "On sequential designs for maximizing the sum of n observations." *Annals of Mathematical Statistics* 27: 1060–74.

Duflo, M. (1996) *Algorithmes Stochastiques*, Berlin: Springer.

Durrett, R. (2004) *Probability: Theory and Examples*, 3rd edn, Belmont, CA: Duxbury Press.

Flache, A. and M. Macy (2002) "Stochastic collusion and the power law of learning." *Journal of Conflict Resolution* 46: 629–53.

Pemantle, R. (1988) "Random processes with reinforcement." Doctoral Dissertation, MIT.

Pemantle, R. (1990) "Nonconvergence to unstable points in urn models and stochastic approximations." *Annals of Probability* 18: 698–712.

Pemantle, R. (2007) "A survey of random processes with reinforcement." *Probability Surveys* 4: 1–79.

Pemantle, R. and B. Skyrms (2001) "Reinforcement schemes may take a long time to exhibit limiting behavior." Preprint.

Pemantle, R. and B. Skyrms (2003) "Time to absorption in discounted reinforcement models." *Stochastic Process Applications* 109: 1–12.

Pemantle, R. and S. Volkov (1999) "Vertex-reinforced random walk on Z has finite range." *Annals of Probability* 27: 1368–88.

Skyrms, B. (2004) *The Stag Hunt and the Evolution of Social Structure*. Cambridge: Cambridge University Press.

Skyrms, B. and R. Pemantle (2000) "A dynamic model of social network formation." *Proceedings of the National Academy of Science*, USA 97: 9340–6.

Taylor, P. and L. Jonker (1978) "Evolutionary stable strategies and game dynamics." *Mathematical Biosciences* 40: 145–6.

Wei, L. and S. Durham (1978) "The randomized play-the-winner rule in medical trials." *Journal of the American Statistical Association* 73: 840–3.

15

Inventing New Signals

with Jason McKenzie Alexander and Sandy L. Zabell

1. Introduction

Sender–receiver signaling games were introduced by Lewis (1969) and in more general form by Crawford and Sobel (1982). Nature picks a state of the world, with some fixed probability, from a set of states. One player, the sender, observes the state and picks a signal from some arbitrary set of signals. (Signals are arbitrary in the sense that they are not assumed to have preexisting meaning or salience.) A receiver observes the signal and chooses one of a set of acts. Payoffs are jointly determined by the state of the world and the act taken. It is interesting to investigate whether some form of adaptive dynamics of evolution or learning can spontaneously generate meaningful signaling.

Recent investigations have demonstrated unexpected complexity in the dynamics of very simple signaling games with strong common interest. Suppose there are the same (finite) number of states, signals and acts, and for each state there is a unique act such that the payoff for both sender and receiver is 1 if the act is chosen and 0 otherwise. (In this case one says there is pure or strong common interest.) A *sender strategy* is a map from states to signals, a *receiver strategy* a map from signals to acts. If we give the act that pays off in a state the same index as the state, we can define a *signaling system equilibrium* as a pair of sender and receiver strategies whose composition maps the i-th state to the i-th act for each index i. This is obviously the most desirable situation to be in.

At the opposite extreme, there are also *complete pooling equilibria*, in which the sender sends the signals with probability independent of the state, the receiver acts with probability independent of the signal received, and consequently the probability of an act being taken does not depend on the state Nature has chosen. Provided there are more than two states, there are also *partial pooling equilibria* in which some but not all states are pooled. (This means that the states and their corresponding acts can be divided into subsets, such that within a subset one has complete pooling: the probability of doing acts in the subset does not depend on states in the subset, and for each state in the subset the sum of the probabilities of the acts in the subset sum to one.)

In these special games, signaling system equilibria are distinguished from an evolutionary point of view by being the unique evolutionarily stable states (Wärneryd 1993). It might then seem plausible that replicator dynamics (Taylor and Jonker 1978) would always lead to a signaling system equilibrium, but this turns out not to be true (Hofbauer and Huttegger 2008; Pawlowitsch 2008). In the special case of 2 states, 2 signals, and 2 acts with pure common interest and where nature chooses states with equal probability, it is true. But if states are not equiprobable, the connected component of *pooling* equilibria has a basin of attraction of positive measure. And with 3 states, 3 signals, and 3 acts, *partial pooling* equilibria have a basin of attraction of positive measure even if nature chooses states with equal probability.

With reinforcement learning, the situation is more complicated. Argiento et al. (2009) consider the basic reinforcement learning scheme of Roth and Erev (1995) and Erev and Roth (1998), applied to 2 state, 2 signal, 2 act signaling games with pure common interest and equiprobable states. It generates a stochastic process as follows. The sender has an urn for each state, each containing two colors (say yellow and blue). When the sender observes a state, she draws a ball from the urn for that state. If she draws a yellow ball she sends signal one; if she draws a blue ball she sends signal two. The receiver has an urn for each signal, each containing two colors (say red and green). Upon receiving a signal, the receiver draws a ball from the urn corresponding to the signal. If the ball is red he does act one; if green, act two. In the case of both sender and receiver, the original balls drawn are then returned. In addition, if the right act for the state is done both sender and receiver also add an extra ball of the same color drawn to the selected urns. This is then repeated. (The initial number of yellow and blue balls in the sender's urns, and the initial number of red and green balls in the receiver's urns can be arbitrary.) Using stochastic approximation theory, Argiento et al.

(2009: Theorem 1.1) prove that with probability one the players converge to a signaling system; if the initial distribution of colors is uniform, each signaling system has equal probability of being selected. Simulation studies suggest this result does not carry over to the case where states are not equiprobable, because players may sometimes converge to pooling.

Pooling is inefficient and undesirable. Why can't the agents simply *invent new signals* to remedy the situation? We would like to have a simple, easily-studied model of such a process. That is to say, we want to move beyond closed models where the theorist fixes the signals, to an open model in which the space of signals itself can evolve. We would like to suggest such an open model here. This involves a kind of hybrid of the Roth–Erev urn process and the Chinese restaurant process—the latter being also known in another guise as the Hoppe–Pólya urn.

2. The Chinese Restaurant Process and the Hoppe–Pólya Urn

Imagine a Chinese restaurant, with an infinite number of tables, each of which can hold an infinite number of guests. People enter one at a time and sit at either an occupied table or an unoccupied one. Imagine there is also a ghost—a phantom guest—who is always sitting at one, otherwise unoccupied, table. The probability that a guest sits down at a table is proportional to the number already at that table, including the phantom guest. (So that if n guests have been seated, the probability of the next guest joining the phantom is $1/(n + 1)$.)

The first person to enter sits at the first unoccupied table, since no one but the phantom guest is there. The phantom guest then moves to an unoccupied table. The second person to enter now has equal probability of either sitting with the first, or sitting at the table with the phantom, resulting in a new occupied table. Should the second person join the first, the third person entering the room has a 2/3 chance of sitting at their table and a 1/3 chance of starting a new occupied one. Should the second person start a new table, the phantom guest moves on, and the third person will join one or the other, or start a third occupied table, all with equal probability. This is the *Chinese Restaurant Process* which has been studied as a problem in abstract probability theory (Aldous 1985; Pitman

1995). It is equivalent to a simple urn model, and this urn scheme can be modified to represent reinforcement learning with invention.

In 1984 Hoppe introduced what he called "Pólya-like Urns" in connection with "neutral" evolution—evolution in the absence of selection pressure. In the classical Pólya urn process, we start with an urn containing various colored balls. Then we proceed as follows: A ball is drawn at random and then returned to the urn with another ball of the same color. All colors are treated in exactly the same way. We can recognize the Pólya urn process as a special case of reinforcement learning in which there is no distinction worth learning—there are no states, no acts, and reinforcement always occurs. It is a standard result that the probabilities in a Pólya urn process (the fraction of each color present) almost surely converge to some random limit. (That is, they are guaranteed to converge to something, but that something can be anything.)

To the Pólya urn, Hoppe (1984) adds a black ball—the *mutator*. The mutator brings new colors into the game. If the black ball is drawn, it is returned to the urn and a ball of an entirely new color is added to the urn. (Hoppe allowed that there might be more than one black ball, corresponding to multiple phantom guests in the Chinese restaurant process. Here, however, we will stick to the simplest case.) The Hoppe–Pólya urn model was meant as a model for neutral evolution, where there are a vast number of potential mutations which convey no selective advantage. (The same urn model has an alternative life in the Bayesian theory of induction, having essentially been invented in 1838 by the logician Augustus de Morgan to deal with the prediction of the emergence of novel categories. It generalizes Bayes–Laplace rules of succession, known to philosophers as Carnap's continuum of inductive methods (Zabell 1992, 2005).)

It is evident that the Hoppe–Pólya urn process and the Chinese Restaurant process are essentially the same process described in two different ways. Hoppe's colors correspond to the tables in the Chinese Restaurant; the mutator ball to the phantom guest. After a finite number of iterations N, the N guests in the restaurant or the N balls in Hoppe's urn (other than the phantom guest or the black ball) are partitioned into some number of categories. The categories are colors for the urn, tables for the restaurant. But the partitions we end up with can be different each time; they depend on the luck of the draw. We have *random partitions of the guests or balls*, each time having a different number of categories, different numbers of individuals in each category, and different individuals filling out the numbers.

The possible patterns that can arise after n guests are seated are described by the *partition vector* $<a_1, \ldots a_n>$ where a_j denotes the number of tables with j guests. For example, if there are four guests sitting at two tables, the pattern of one table with one guest and one table with three guests means that $a_1 = 1$ (the one table with one guest), $a_2 = 0$ (there are no tables with two guests), $a_3 = 1$ (the one table with three guests), and $a_4 = 0$ (there are no tables with four guests); this corresponds to the partition vector $<1,0,1,0>$.

Note that there are four ways of realizing this pattern and that all four are equally likely. If the two tables are labeled A and B, and the four guests arriving in order are labeled 1, 2, 3, 4, then the four patterns and their probabilities are (the first guest always sits at the first table):

Table A	Table B	Probability
1, 2, 3	4	$1 \times 1/2 \times 2/3 \times 1/4 = 1/12$
1, 2, 4	3	$1 \times 1/2 \times 1/3 \times 2/4 = 1/12$
1, 3, 4	2	$1 \times 1/2 \times 1/3 \times 2/4 = 1/12$
1	2, 3, 4	$1 \times 1/2 \times 1/3 \times 2/4 = 1/12$

It is in fact generally true of the process that all realizations of a given partition vector are equally likely to occur. All that affects the probability is a specification of the number of tables that have a given number of guests. The fact that any arrangement of guests with the same partition vector has the same probability is called *partition exchangeability*, and it is the key to mathematical analysis of the process.

There are explicit formulas to calculate probabilities and expectations of classes of outcomes after a finite number of trials. The expected number of categories—of colors of ball in Hoppe's urn or the expected number of tables in the Chinese restaurant—will be of particular interest to us, because the number of colors in a sender's urn will correspond to number of signals in use. This is given by a very simple formula: after N iterations, the expected number is $\sum_{(i=1 \text{ to } N)} 1/i$, the N-th partial sum of the harmonic series, which grows logarithmically in N. Results are plotted in Figure 15.1.

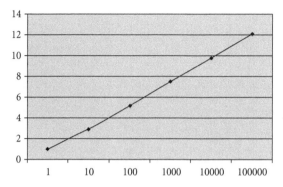

Figure 15.1 Expected number of categories

Although it is known that with probability one the limiting number of categories is infinite, for even quite large numbers of trials the expected number of categories is relatively modest.

There is something else that we would like to emphasize. For a given number of categories, the distribution of trials among those categories is not uniform. We can illustrate this with a simple example. Suppose we have ten trials and the number of categories turns out to be two (two colors of ball, two tables in the restaurant), something that happens about 28 per cent of the time. This can be realized in five different ways as a partition of 10: $5 + 5$, $4 + 6$, $3 + 7$, $2 + 8$, $1 + 9$. (These correspond to the partition vectors: $a_5 = 2$ and $a_j = 0$ otherwise, $a_4 = a_6 = 1$ and $a_j = 0$ otherwise, $a_3 = a_7 = 1$ and $a_j = 0$ otherwise, $a_2 = a_8 = 1$ and $a_j = 0$ otherwise, and $a_1 = a_9 = 1$ and $a_j = 0$ otherwise, respectively.) There is a simple way of calculating the probability of each—the *Ewens sampling formula*—which gives the probability of a partition vector for n draws:

$$\Pr<a_1, \ldots, a_n> = \prod_{j=1 \text{ to } n} 1/[(j^{a_j})(a_j!)]$$

The results in the case of our example are graphed in Figure 15.2.

The more unequal a division is between the categories, the more likely it is to occur. Some colors are numerous, some are rare. Some tables are much fuller than others. This can be seen as the result of a kind of *preferential attachment* process. In the Chinese restaurant, fuller tables are more likely to attract new guests. This generates a power-law distribution, similar to those that are ubiquitous in word frequencies in natural language and elsewhere (Zipf 1932).

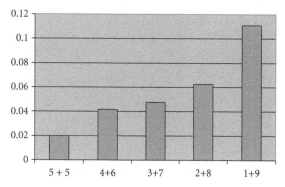

Figure 15.2 Probability of partitions of 10 into two categories

3. Reinforcement with Invention

We remarked that the Pólya urn process can be thought of as reinforcement learning when there is no distinction worth learning—all choices (colors) are reinforced equally. The Hoppe–Pólya urn, then, is a model which adds useless invention to useless learning. That was its original motivation, where different alleles confer no selective advantage.

If we modify the Pólya urn by adding differential reinforcement—where choices are reinforced according to different payoffs—we get the basic Roth–Erev model of reinforcement learning, used by Argiento et al (2009). If we modify the Hoppe–Pólya model by adding differential reinforcement, then—as discussed below—we get reinforcement learning capable of invention.

4. Inventing New Signals

We use the Hoppe–Pólya urn as a basis for a model of inventing new signals in signaling games. For each state of the world, the sender has an additional choice: *send a new signal*. A new signal is always available. The sender can either send one of the existing signals or send a new one. Receivers always pay attention to a new signal. (A new signal means a new signal that is noticed, failures being taken account of by making the probability of a successful new signal smaller than one.) Receivers, when confronted with a new signal, just act at random. We equip them with equal initial propensities for the acts.

Before successful invention

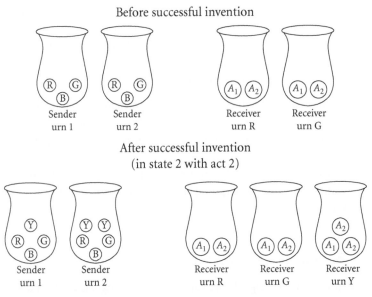

Figure 15.3 R denotes a red ball, G green, B black. In State 2 a black ball is drawn, Act 2 is tried and is successful. A yellow ball is added to both senders' urns and a reinforcement yellow ball is added to the urn for State 2. The receiver adds an urn for the signal yellow, and adds an extra ball to that urn for Act 2

Now we need to specify exactly how learning proceeds. Nature chooses a state and the sender either chooses a new signal, or one of the old signals. If there is no new signal the model works just as before with basic Roth–Erev reinforcement. If a new signal is introduced, it either leads to a successful action or not. When there is no success, the system returns to the state it was in before the experiment with a new signal was tried.

But if the new signal leads to a successful action, both sender and receiver are reinforced. The reinforcement now consists of the sender's increased propensity to send the signal in the state in which it was just sent, and the receiver keeping track of the successful acts when receiving the new signal. In terms of the urn model, in the case of a successful act, the receiver activates an urn for the signal, with one ball for each act, and a second ball for the successful act. The sender now considers the new signal not only in the state in which it was tried out, but also considers it a possibility in other states. So, in terms of the urn model, a ball for the new signal is added to each sender's urn, as well as a reinforcement ball added to the urn for the state that just occurred. The new signal has now established itself. We have moved from a Lewis signaling game with N signals to one with $N + 1$ signals. See Figure 15.3.

In summary, one of three things can happen:

1. No new signal tried, and the game is unchanged. Reinforcement proceeds as in a game with a fixed number of signals.
2. A new signal is tried but without success, and the game is unchanged.
3. A new signal is tried with success, and the game changes from one with n states, m signals, and p acts to one with n states, $m + 1$ signals, p acts.

5. Starting with Nothing

If we can invent new signals, we can start with no signals at all, and see how the process evolves. We can expect that—like the simple Hoppe-Pólya urn—the limiting number of different signals will be infinite. The appendix gives a proof for the case of m states having unequal probability and n acts. Starting with no signals, the limiting number of different types of signals is almost surely infinite, and each signal is almost surely sent an infinite number of times.

But if we run a large finite number of iterations, we do not expect a comparably large of signals. For learning with invention, we do not have anything like the stochastic approximation theory analysis of Argiento et al. (2009) to prove this rigorously, and such an analysis looks very hard to come by. We therefore proceed with a preliminary investigation by simulation.

Consider the 3 state, 3 act Lewis signaling game with states equally probable. As before, we have strong common interest—exactly one act is right for each state. In simulations of our model of invention, starting with no signals at all, the number of signals after 100,000 iterations ranged from 5 to 25. A histogram of the final number of signals in 1,000 trials is shown in Figure 15.4. This behavior is close to that which would be expected from a pure Chinese Restaurant process.

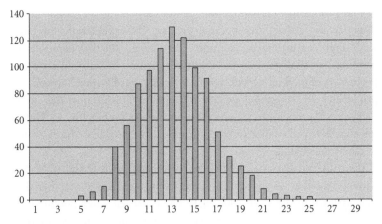

Figure 15.4 Number of signals after 100,000 iterations of reinforcement with invention. Frequency in 1000 trials

6. Avoiding Pooling Traps

In a version of this game with the number of signals fixed at 3, reinforcement learning sometimes falls into a partial pooling equilibrium. In simulations of basic Roth–Erev reinforcement learning with initial propensities of 1, 9.6 percent of the trials led to imperfect information transmission. (Barrett 2007; 1000 trials, 1,000,000 iterations of learning per trial). In these cases the average payoff approaches 2/3 and the players appear to approach a partial pooling equilibrium. (In the remaining trials the average payoff was close to 1.) Using reinforcement with invention, starting with no signals, 1,000 trials *all* ended up with efficient signaling (using a success rate of at least 99 per cent after one million trials as a proxy for efficient signaling). Signalers went beyond inventing the three requisite signals. Lots of synonyms were created. By inventing more signals, they avoided the traps of partial pooling equilibria.

In the game with 2 states, 2 acts, and the number of signals fixed at 2, if the states had unequal probabilities agents sometimes fell into a complete pooling equilibrium—in which no information at all is transmitted and the average payoff is 1/2. In such an equilibrium the receiver would simply do the act suited for the most probable state and ignore the signal, and the sender would send signals with probabilities that were insensitive to the state.

The probability of falling into complete pooling increased as the disparity in probabilities became greater. Our simulations of basic Roth–Erev reinforcement learning, 1,000 trials, 100,000 iterations of learning per trial, give us the following picture. When one state has probability .6, failure of information transfer hardly ever happens. At probability .7 it happens 5 percent of the time. This number rises to 22 percent for probability .8, and 38 percent for probability .9. Highly unequal state probabilities appear to be a major obstacle to the evolution of efficient signaling.

If we take an extreme case in which one state has probability .9, start with no signals at all, and let the players invent signals as above, then they reliably learn to signal. In 1000 trials they never fell into a pooling trap; they always learned a signaling system (again using a success rate of at least 99 percent after one million trials as a proxy for efficient signaling). The invention of new signals makes efficient signaling a much more robust phenomenon.

7. Causes of Efficiency

The ability to avoid partial pooling traps and evolve efficient signaling might be due to two different mechanisms, alone or in combination. It might be the case that an excess of signals might, by itself, make partial pooling much less likely. If this is the case, then starting with a fixed number of signals larger than the number of states (invention disabled) would make it less likely that individuals would get near partial pooling. Or individuals falling into a partial pooling equilibrium might be able to invent themselves out of it. If this were the case, then we should find that if we start the process with invention near partial pooling, new signals enable the evolution to efficient signaling. In fact, both mechanisms can operate.

Consider the case of two states, signals and acts, in which state 1 has probability .9, and state 2 probability .1. We will initialize the system near a pooling equilibrium, and run learning with invention. The degree of entrenchment of the pooling equilibrium can be varied by changing the initial weights: with more balls in the sender's and receiver's urn, it becomes more difficult to wander away from pooling.

We start the sender with $(\frac{1}{2}n) + 1$ balls for signal 1 and $(\frac{1}{2}n) + 1$ balls for signal 2 in the urn for each state; the receiver with $n + 1$ balls for act 1

Table 15.1 The effect of extra signals on efficient signaling

Number of initial signals	% of pooling	% of signaling
2	38.1	61.9
3	12.0	88.0
4	4.5	95.5
5	1.7	98.3
6	0.5	99.5
7	0.5	99.5
8	0.2	99.8
9	0.0	100.0

and 1 ball for act 2 in her urn for each signal. The higher the entrenchment parameter, n, the more difficult it is for a new signal to become established. For $n = 10, 100, 1,000, 10,000$ learning with inventions always converged to a signaling system equilibrium (1,000 trials, 1,000,000 iterations per trial). For comparison, with no invention learning always converged to the pooling equilibrium with $n > 100$. (The results were similar if the system was initialized near the pooling equilibrium where signal 1 was always sent.)

Even in simulations with 1,000,000 learning steps, invention produces only a modest number of signals. So we might just start the process with extra signals already there. We now keep the initial weights as in the original setup and vary the initial number of signals. Extra signals promote learning to signal efficiently, as shown in Table 15.1.

It appears that both (i) excess initial signals make it less likely to fall into a pooling equilibrium and (ii) invention of new signals allows escape from the vicinity of a pooling equilibrium.

8. Synonyms

Let us return to signaling with invention. Typically we get efficient signaling with lots of synonyms. How much work are the synonyms doing? Consider the simulation illustrated in Table 15.2 of $3 \times 3 \times 3$ signaling, starting with no signals and proceeding with 100,000 iterations of learning with invention.

Notice that a few of the signals (shown in boldface) are doing most of the work. In state 1, signal 5 is sent 87 percent of the time. Signals 1 and 2

Table 15.2 In invention, only a few synonyms are used to achieve efficiency

signal 1	probabilities in states 0,1,2	0.000, **0.716**, 0.000
signal 2	probabilities in states 0,1,2	0.000, **0.281**, 0.000
signal 3	probabilities in states 0,1,2	0.096, 0.000, 0.001
signal 4	probabilities in states 0,1,2	0.009, 0.000, 0.000
signal 5	probabilities in states 0,1,2	**0.868**, 0.000, 0.000
signal 6	probabilities in states 0,1,2	0.000, 0.000, **0.810**
signal 7	probabilities in states 0,1,2	0.024, 0.000, 0.000
signal 8	probabilities in states 0,1,2	0.000, 0.000, **0.143**
signal 9	probabilities in states 0,1,2	0.000, 0.000, 0.044
signal 10	probabilities in states 0,1,2	0.000, 0.000, 0.000
signal 11	probabilities in states 0,1,2	0.000, 0.000, 0.000
signal 12	probabilities in states 0,1,2	0.001, 0.000, 0.000
signal 13	probabilities in states 0,1,2	0.000, 0.000, 0.000

function as significant synonyms for state 2, being sent more that 99.5 percent of the time. Signals 6 and 8 are the major synonyms for state 3. (All of these signals are highly reinforced on the receiver side.) The pattern is fairly typical (in 1,000 trials). Very often, many of the signals that have been invented end up being little used.

This is just what we should expect from what we know about the Hoppe–Pólya urn. Even without any selective advantage, the distribution of reinforcements among categories tends to highly unequal, as was illustrated in Figure 5.2. Might not infrequently used signals simply fall out of use entirely? At present we do not know, only simulations that suggest they do not fade away.

9. Noisy Forgetting

Nature forgets things by having individuals die. Some strategies (phenotypes) simply go extinct. This cannot really happen in the replicator dynamics—an idealization where unsuccessful types get increasingly rare but never actually vanish. And it cannot happen in Roth–Erev reinforcement where unsuccessful acts are dealt with in much the same way.

Evolution in a finite population is different. In the models of Sebastian Shreiber (2001), a finite population of different phenotypes is modeled as an urn of balls of different colors. Successful reproduction of a phenotype corresponds to the addition of balls of the same color. So far this is identical to the basic model of reinforcement learning. But in Schreiber's models individuals also die. We transpose this idea to learning dynamics to get a model of reinforcement learning with noisy forgetting.

For individual learning, this model may be more realistic than the usual model of geometrical discounting. (Suggested in Roth and Erev as a modification of the basic model to incorporate forgetting or "recency" effects.) That model, which discounts the past by keeping some fixed fraction of each ball at each update, may be best suited for aggregate learning—where individual fluctuations are averaged out. But individual learning is noisy, and it is worth looking at an urn model of individual reinforcement with noisy forgetting.

10. Inventing and Forgetting Signals

We can put together these ideas to get learning with invention and with noisy forgetting, and apply it to signaling. It is just like the model of inventing new signals except for the random dying-out of old reinforcement, implemented by random removal of balls from the sender's urns.

The idea may be implemented in various ways. Here is one. With some probability, nature picks an urn at random and removes a colored ball at random. (The probability is the forgetting rate, and we can vary it to see what happens.) Call this *Forgetting A*. Here is another. Nature picks an urn at random, picks a color represented in that urn at random, and removes a ball of that color. Call this *Forgetting B*.

Now it is possible that the number of balls of one color, or even balls of all colors could hit zero in a sender's urn. Should we allow this to happen, as long as the color (the signal) is represented in other urns for other states? Here is another choice to be made. But in the simulations we are about to report, we never hit a zero.

We simulated *invention* with *No Forgetting, Forgetting A, Forgetting B,* starting with no signals, for the number of states (=acts) being 3, 4, 5. States are equiprobable and the forgetting rate is .3 for all simulations. A run consists of 1,000,000 iterations of the learning process, and each entry is the average of 1,000 runs. The results are shown in Table 15.3.

Relative to *No Forgetting, Forgetting B* is highly effective in pruning unused signals. (However, as noted before, even with no forgetting the number of excess signals after a million trials is not that large.)

Table 15.3 Using forgetting to prune unneeded signals

	Average number of signals remaining		
	No forgetting	Forgetting A	Forgetting B
3 states	16.276	19.879	3.016
4 states	17.491	21.079	4.005
5 states	18.752	22.686	4.982
6 states	20.097	24.069	5.975
7 states	21.336	25.820	6.960
8 states	22.661	27.140	7.941
9 states	23.815	28.684	8.929
10 states	24.925	30.663	9.928

In contrast, *Forgetting A* does not help at all and even appears to be detrimental. This difference is real: even after a run of only 1,000 iterations of the learning process and 100 runs for both *No Forgetting* and *Forgetting A*, the difference in the average of the number of signals (10.90—9.18 = 1.72) is highly statistically significant. (Using the Welch—that is, unpooled—test, $t = 4.4579$ on 197 degrees of freedom, with a p-value of 1.388×10^{-5}; the corresponding 95 percent confidence interval for the difference is 0.96 to 2.48. The difference remains highly significant even using a nonparametric procedure: the Wilcoxon rank sum test gives an even smaller p-value of 5.061×10^{-6}.)

Why is there a difference between *No Forgetting* and *Forgetting A*? A plot of the average number of signals for the two cases as one goes from 1 to 1,000 iterations, as shown in Figure 15.5, is suggestive:

This behavior does not appear to depend on any special properties of the equiprobable state case just discussed: if one does 1,000 iterations of *No Forgetting* and *Forgetting A* for 100 randomly chosen state probabilities instead, a similar graph arises, as shown in Figure 15.6.

We hypothesize that one important reason for why *No Forgetting* does as well as it does is that it is relatively easy for a successful signal to lock in, and the number of the color balls that represent it to grow quickly. With *Forgetting A*, the more balls there are of a given color, the more likely it is that one of that color will be thrown away. This tends to prevent lock-in, make reinforcement of a successful color less likely, and hence increase the variety of colored balls in the urn.

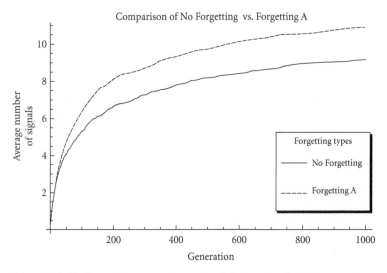

Figure 15.5 *No Forgetting* versus *Forgetting A*, 100 runs of 1,000 generations, states equiprobable

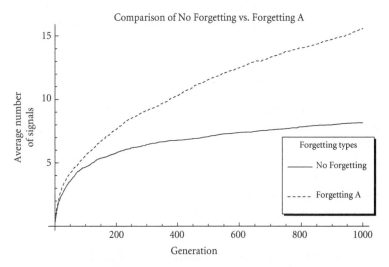

Figure 15.6 *No Forgetting* versus *Forgetting A*, 100 runs of 1,000 generations. Initial state probabilities selected at random from a uniform distribution over the simplex. The same initial conditions were used for the *n*th run of each type of forgetting

11. Related Work: Infinite Numbers of States or Acts

Bhaskar (1998) analyzes an evolutionary model with an infinite number of signals. There is noisy, cost-free pre-play communication added to a base game. Because of the infinite number of signals, there must be signals with arbitrarily low probability of use. Then, because of special properties of the model, there are always signals that are as good as new in functioning as a "secret handshake" (Robson 1989). The "secret handshake" destabilizes inefficient equilibria. This suggests investigation of learning with inventing new signals in the setting of cost-free pre-play communication.

Worden and Levin (2007) analyze a model in which there are an infinite number of potential acts. Trying an unused act changes the game. It is assumed that players can only try an unused act that is "close" to a used act in that its payoff consequences are epsilon-close to those of one of the old acts. The game can then change slowly so that a Prisoner's Dilemma can evolve into a non-dilemma. Formally, we could extend our model to include inventing new acts by giving the receiver a Hoppe–Pólya urn. The question is how to extend the payoff matrix. Worden and Levin supply one sort of answer. Others might be of interest, depending on the context of application.

12. Conclusion

We move from signaling games with a closed set of signals to more realistic models in which the number of signals can change. New signals can be invented, so the number of signals can grow. Little used signals may be forgotten, so the number of signals can shrink. A full dynamical analysis of these models is not available, but simulations suggest that these open models are more conducive to the evolution of efficient signaling than previous closed models.

Appendix: Infinite Number of Signals

This appendix recapitulates some definitions in more mathematical form, and provides proofs of some statements.

A.1. Signaling Equilibria

Suppose there are s states, s acts, and t signals. A sender strategy is an $s \times t$ stochastic matrix $A = (a_{ij})$; a_{ij} is the probability that the sender sends signal j given state i. A receiver strategy is a $t \times s$ stochastic matrix $B = (b_{jk})$; b_{jk} is the probability that the receiver chooses act k given he receives signal j. In this case the pair of strategies (A, B) is said to be a signaling system. Note that the matrix product $C = AB$ is an $s \times s$ stochastic matrix; if $C = (c_{ik})$, then c_{ij} is the probability that the receiver chooses act k given state i.

If π_i is the probability that nature chooses state i, then $\sum_i \pi_i c_{ii}$ is the probability that a correct state-act pairing occurs. This is the same as the expected payoff for the signaling system (given the assumed payoff structure). By an *equilibrium* we mean a Nash equilibrium: that is, neither the sender nor the receiver can increase the common expected payoff by one changing their strategy while the other does not. A *signaling system equilibrium* is a pair of strategies (A, B) such that $C = I$, the identity matrix. In this case $t \geq s$ (there must be at least as many signals as states), and if $a_{ij} > 0$, then $b_{ij} = 1$.

A group of states is said to be *pooled* in a sender strategy if for every signal the probability of that signal being sent is the same for every state in the group. A *complete pooling equilibrium*, an equilibrium in which all the states are pooled; that is, it is a sender strategy A all of whose rows are the same (the probability of sending a particular signal is independent of the state), and therefore the rows of C are also equal (the probability of choosing a particular act is also independent of the state). A *partial pooling equilibrium* is an equilibrium in which some but not all of the states are pooled.

A.2. The Ewens Sampling Formula

The Chinese restaurant process/Hoppe–Pólya urn provides a stochastic mechanism for generating the class of *random partitions* described by the *Ewens sampling formula*. Given n objects divided into t classes (each class containing at least one object), with n_i in the ith class, let a_j $1 \leq j \leq n$, record the number of classes containing j objects (that is, a_j is the number of $n_i = j$). Then $<n_1, \ldots, n_t>$ is the vector of (class) *frequencies* and $<a_1, \ldots, a_n>$ the corresponding *partition vector*. Clearly one has

$$\sum_{i=1}^{t} n_i = \sum_{j=1}^{n} j a_j = n, \sum_{j=1}^{n} a_j = t$$

In the case of the Chinese restaurant process the number of classes present after n objects are generated is itself random. In the most general case there is a fre parameter θ, and if $S_n(\theta) = \theta(\theta + 1)(\theta + 2) \ldots (\theta + n - 1)$, then the probability of seeing a particular partition vector $<a_1, \ldots, a_n>$ is given by the Ewens formula

$$P(<a_1, \ldots, a_n>) = \frac{n! \theta \sum_{j=1}^{n} a_j}{\prod_{j=1}^{n} j^{a_j} a_j! S_n(\theta)}$$

If $\theta = 1$, then this reduces to the formula given earlier in this paper. If instead $\theta = 2$, say, then this would correspond to there being two black mutator balls (or phantom guests in the CRP version), rather than just one; and similarly for θ any positive integer. (But the formula makes sense for any positive value of θ integral or not; this corresponds to giving the black ball a relative weight of θ versus the unit weight given every colored ball.)

There is a rich body of theory that has arisen over the last four decades concerning the properties of random partitions whose stochastic structure is described by the Ewens formula.

If T_n denotes the (random) number of different type or colors (other than black) in the urn after n draws, then one can show that its expected value is

$$E[T_n] = \theta \left(\frac{1}{\theta} + \frac{1}{\theta + 1} + \ldots + \frac{1}{\theta + n - 1} \right).$$

If $\theta = 1$, this reduces to the nth partial sum of the harmonic series,

$$\sum_{k=1}^{n} \tfrac{1}{k} = \text{In } n + \gamma + \epsilon_n;$$

here $\gamma = 0.57721 \ldots$ is Euler's constant, $\epsilon_n \sim 1/2n$ (that is, $\lim_{n \to \infty} 2n\epsilon_n = 1$).

A.3. Two Limit Theorems

In the following we derive some of the properties of the signaling with invention process described in the section on inventing new signals.

A.3.1 The number of different colors diverges

First some notation. Let $\mathcal{F}_n$ be history of the process up to time n (the nth trial). Let A_n denote the event that on the nth trial a new signal is tried with success. Let ω denote the entire infinite history of a specific realization of our reinforcement process. (So if, instead, one were considering the process of tossing a coin an infinite number of times, ω would represent a specific infinite sequence of heads and tails.) Let

$$P(A_n \mid \mathcal{F}_{n-1})$$

denote the conditional probability that A_n occurs, given the past history of the process up to time $n-1$. This is not a number, but a random quantity, since it depends on the realization ω, which is random. Finally, let

$$P(A_n \mid \mathcal{F}_{n-1})(\omega)$$

denote this conditional probability for a specific history or realization ω; this is a number.

By the martingale generalization of the second Borel–Cantelli lemma (see e.g., Durrett 1996: 249), one has

$$\{A_n \text{ i.o.}\} = \left\{ \sum_{n=1}^{\infty} P(A_n \mid \mathcal{F}_{n-1}) = \infty \right\} \text{ almost surely.}$$

That is, consider the following two events. The first is

$$\{\omega : \omega \in A_n \text{ infinitely often}\};$$

the second event is

$$\left\{ \omega : \sum_{n=1}^{\infty} P(A_n \mid \mathcal{F}_{n-1})(\omega) = \infty \right\}.$$

The assertion is that the two events are the same, up to a set of probability zero.

We claim that A_n occurs infinitely often with probability one; that is, $P(\{A_n \text{ i.o}\}) = 1$.

By the version of the Borel–Cantelli lemma just cited, it suffices to show that

$$P\left(\left\{\omega : \sum_{n=1}^{\infty} P(A_n|\mathcal{F}_{n-1})(\omega) = \infty\right\}\right) = 1.$$

In fact we show more: that

$$\sum_{n=1}^{\infty} P(A_n|\mathcal{F}_{n-1})(\omega) = \infty$$

for *every* history ω.

To see this, suppose that there are k states, and that the sender selects these with probabilities p_j, $1 \leq j \leq k$. Suppose that initially there is just one ball, the black ball, in each of the k urns. Then at each stage there are $a_j \geq 1$ balls in each urn, one black and the remaining $a_j - 1$ some variety of colors. The probability that at stage n a new signal is generated and successfully used depends on the values of $a_1, \ldots, a_k$ at the start of stage n (that is, before selection takes place), and is

$$\sum_{j=1}^{k} p_j\left(\frac{1}{a_j}\right)\left(\frac{1}{k}\right).$$

(That is, you pick the jth urn with probability p_j, you pick the one ball out of the a_j that is black, and there is a one chance in k that the receiver chooses the correct act.)

Now use the generalized *harmonic mean–arithmetic mean inequality* (see, e.g., Hordy et al. 1988); this tells us that for $a_j > 0$, one has

$$\frac{1}{\sum_{j=1}^{k}\left(\frac{p_j}{a_j}\right)} \leq \sum_{j=1}^{k} p_j a_j.$$

Further, the total number of balls in the state urns, $a_1 + \ldots + a_k$, is greatest when a black ball has been selected every time and the receiver chooses the right act (since then one adds $k + 1$ balls of a new color at each stage rather than just one). Thus at the start of stage n one has

$$a_1 + \ldots + a_k \le (k+1)(n-1) + k$$

Thus if $p^* = \max\{p_j, 1 \le j \le k\}$, it is apparent that

$$\sum_{j=1}^{k} p_j a_j \le p^*(a_1 + \ldots + a_k) \le p^*[(k+1)n - 1].$$

Putting this all together gives us that the probability that at stage n new signal is generated and successfully used is

$$\frac{1}{k} \sum_{j=1}^{k} \left(\frac{p_j}{a_j}\right) \ge \frac{1}{kp^*} \frac{1}{(k+1)n - 1}.$$

It follows that

$$\sum_{n=1}^{\infty} P(A_n | \mathcal{F}_{n-1})(\omega) \ge \frac{1}{kp^*} \sum_{n=1}^{\infty} \frac{1}{(k+1)n - 1} > \frac{1}{k(k+1)p^*} \sum_{n=1}^{\infty} \frac{1}{n} = \infty,$$

using the well-known fact that the harmonic series diverges.

A.3.2 The number of balls of a given color diverges

Suppose a color has been established. The above proof can be easily modified to show that the number of balls of a given color in all the state urns, say green, tends to infinity almost surely as n tends to infinity. Let A_n now denote the event that on the nth trial green is selected and reinforced; and $p_* = \min\{p_j\ 1 \le j \le k\} > 0$. Then in the receiver's green urn there must be at least one state that is represented at least $1/k$ of the time (since there are k states and the sum of the fractions must sum to one). Let a_{j^*} denote the index of this state. (Note this state can vary with n.) Then one has

$$P(A_n | \mathcal{F}_{n-1}) \ge p^* \frac{1}{a_{j^*}} \left(\frac{1}{k}\right) \ge \frac{p_*}{k[(k+1)n - 1]},$$

and this again diverges. (This bound is obviously quite crude.)

It follows as a simple corollary that the signal corresponding to a given color is necessarily sent (almost surely) an infinite number of times (even though each time the signal is sent its color may or may not be reinforced).

References

Aldous, D. (1985) "Exchangeability and Related Topics" In *l'École d'été de probabilités de Saint-Flour, XIII–1983* 1–198. Berlin: Springer.

Argiento, R., R. Pemantle, B. Skyrms, and S. Volkov (2009) "Learning to Signal: Analysis of a Micro-Level Reinforcement Model." *Stochastic Processes and their Applications* 119: 373–90.

Barrett, J. A. (2007) "Dynamic Partitioning and the Conventionality of Kinds." *Philosophy of Science* 74: 526–46.

Bhaskar, V. (1998) "Noisy Communication and the Evolution of Cooperation." *Journal of Economic Theory* 82: 110–31.

Crawford, V. and J. Sobel (1982) "Strategic Information Transmission." *Econometrica* 50: 1431–51.

Durrett, R. (1996) *Probability: Theory and Examples.* 2nd edn. Belmont, CA: Duxbury Press.

Erev, I. and A. E. Roth (1998) "Predicting how people play games: reinforcement learning in experimental games with unique mixed-strategy equilibria." *The American Economic Review* 88: 848–81.

Hardy, G. H., J. E. Littlewood, and G. Pólya (1988) *Inequalities,* 2nd edn. Cambridge: Cambridge University Press.

Hofbauer, J. and S. Huttegger (2008) "Feasibility of Communication in Binary Signaling Games." *Journal of Theoretical Biology* 254: 843–9.

Hoppe, F. M. (1984) "Pólya-like Urns and the Ewens Sampling Formula." *Journal of Mathematical Biology* 20: 91–4.

Lewis, D. (1969) *Convention.* Cambridge, MA: Harvard University Press.

Pawlowitsch, C. (2008) "Why Evolution Does Not Always Lead to an Optimal Signaling System." *Games and Economic Behavior* 63: 203–26.

Pitman, J. (1995) "Exchangeable and Partially Exchangeable Random Partitions", *Probability Theory and Related Fields* 102: 145–58.

Robson, A. (1989) "Efficiency in Evolutionary Games: Darwin, Nash and the Secret Handshake" *Journal of Theoretical Biology* 144: 379–96.

Roth, A. and I. Erev (1995) "Learning in Extensive Form Games: Experimental Data and Simple Dynamical Models in the Intermediate Term." *Games and Economic Behavior* 8: 164–212.

Shreiber, S. (2001) "Urn Models, Replicator Processes and Random Genetic Drift." *SIAM Journal on Applied Mathematics* 61: 2148–67.

Taylor, P. D. and L. B. Jonker (1978) "Evolutionarily Stable Strategies and Game Dynamics." *Mathematical Biosciences* 40: 145–56.

Wärneryd, K. (1993) "Cheap Talk, Coordination, and Evolutionary Stability." *Games and Economic Behavior* 5: 532–46.

Worden, L. and S. Levin (2007) "Evolutionary Escape from the Prisoner's Dilemma." *Journal of Theoretical Biology* 245: 411–22.

Zabell, S. (1992) "Predicting the Unpredictable." *Synthese* 90: 205–32.

Zabell, S. (2005) *Symmetry and Its Discontents: Essays in the History of Inductive Probability.* Cambridge: Cambridge University Press.

Zipf, G. (1932) *Selective Studies and the Principle of Relative Frequency in Language.* Cambridge, MA: Harvard University Press.

16

Signals, Evolution, and the Explanatory Power of Transient Information

Pre-play signals that cost nothing are sometimes thought to be of no significance in interactions which are not games of pure common interest. We investigate the effect of pre-play signals in an evolutionary setting for Assurance, or Stag Hunt, games and for a Bargaining game. The evolutionary game with signals is found to have dramatically different dynamics from the same game without signals. Signals change stability properties of equilibria in the base game, create new polymorphic equilibria, and change the basins of attraction of equilibria in the base game. Signals carry information at equilibrium in the case of the new polymorphic equilibria, but transient information is the basis for large changes in the magnitude of basins of attraction of equilibria in the base game. These phenomena exemplify new and important differences between evolutionary game theory and game theory based on rational choice.

1. Introduction

Can signals that cost nothing to send have any impact on strategic interaction? Folk wisdom exhibits a certain skepticism: "Talk is cheap." "Put your money where your mouth is." Diplomats are ready to discount signals with no real impact on payoffs. Evolutionary biologists emphasize signals that are too costly to fake (see Zahavi 1975; Grafen 1990; Zahavi and Zahavi 1997). Game theorists know that costless signals open up the possibility of "babbling equilibria," where senders send signals uncorrelated with their types and receivers ignore the signals. Can cost-less signals have any explanatory power at all?

They can in certain kinds of benign interaction where it is in the common interest for signaling to succeed. In fact, even with potential signals that initially have no meaning, it is possible for signals to spontaneously acquire meaning under standard evolutionary dynamics and for such meaningful signaling to constitute an evolutionarily stable strategy. Such spontaneous emergence of signaling systems is often to be expected in strategic interactions such as the sender–receiver games introduced by David Lewis (1969)[1] to give a game-theoretic account of meaning. Signals acquire the ability to transmit information as a result of the evolutionary process. That this is so must be counted as one of the great successes of evolutionary game theory.

But what about interactions where the interests of the parties involved are not so perfectly aligned? In a bargaining game, one strategy may have no interest in whether signaling succeeds or not. In an assurance game one strategy may have an interest in communicating misinformation. How will the evolutionary dynamics of such games be affected if we add a round of costless pre-play signaling, allowing each player to condition her act on the signal received from the other player? It would not be implausible to expect that in such situations costless signals would have little of no effect. Talk is cheap.

In fact, "cheap talk" signaling has a dramatic effect on the evolutionary dynamics of such interactions. That this is so should warn us against quick arguments against the effectiveness of costless signaling. Why it is so illustrates some subtleties of the role of information in the evolutionary process.

Section 2 briefly reviews the evolutionary dynamics in Sender–Receiver Games. Section 3 introduces the embedding of a two-person game in a larger "cheap-talk" game with a round of pre-play costless signaling by both players. Section 4 discusses the effect of cheap talk on the evolutionary dynamics of an assurance game, where rational choice theory predicts that it should have no effect. Section 5 discusses the effect of cheap talk on the evolutionary dynamics of a bargaining game. Section 6 concludes.

---

[1] A more general sender–receiver was introduced and analyzed in Crawford and Sobel (1982).

2. Evolution of Meaning in Sender–Receiver Games

In *Convention*, David Lewis introduced Sender–Receiver games to illustrate a game-theoretic account of conventions of meaning. One player, the sender, has private information about the true state of the world. The other player, the receiver, must choose an act whose payoff depends on the state of the world. The sender has signals available that she can send to the receiver, but they have no exogenously specified meaning. In the model of Lewis there are exactly the same number of states, signals, and acts. In each state of the world there is a unique act that gives both players a payoff of 1, all other acts giving a payoff of 0.

A sender's strategy maps states of the world onto signals. A receiver's strategy maps signals onto acts. There are many Nash equilibria in this game, including ones in which the sender ignores the state of the world and always sends the same signal and the receiver ignores the signal and always chooses the same act. An equilibrium where players always get things right and achieve a payoff of 1 is called a *Signaling System Equilibrium* by Lewis. There are multiple signaling system equilibria. This makes Lewis' point that the meaning of the signals, if they have meaning, is conventional—it depends on which signaling system equilibrium the players exemplify.

From the point of evolutionary game theory there are two striking facts about Lewis signaling games. The first has to do with the evolutionarily stable strategies. Recall that a strategy, s, is *evolutionarily stable* if, for any alternative strategy, m, s does better played against itself than m does, or if they do equally well against s, then s does better against m than m does. (If the latter condition is weakened from "s does better against m than m does" to "s does at least as well against m as m does" the strategy is said to be *neutrally stable*.)

In the evolutionary model for a Lewis game, a player may find herself either in the role of sender or receiver. Her strategy is a pair <sender's strategy, receiver's strategy> from the original Lewis game. Although Lewis sender–receiver games have many equilibria other than signaling system equilibria, the signaling system equilibria coincide with the evolutionarily stable strategies in this evolutionary model.[2]

----

[2] The situation is somewhat complicated if the Lewis model is modified to allow the number of messages to exceed the number of states.

The second fact is more powerful. In a standard model of evolutionary dynamics, the replicator dynamics, the signaling system equilibria are attractors whose joint basin of attraction covers almost all of the possible population proportions. This is shown in simulations, where some signaling system or other *always* goes to fixation. In simplified sender–receiver games, it can be shown analytically (Skyrms 1999). The analytical proof generalizes from the replicator dynamic to a broad class of adaptive dynamics, giving a demonstration of robustness of the result.

3. Evolutionary Games with Cheap Talk

Given the effectiveness of adaptive dynamics in investing signals with meaning, it might be interesting to examine the coevolution of strategy in a game and of meaning in pre-play signals preceding that game. In particular, consider a two-person symmetric game. We imbed this in a cheap talk game by introducing a set of signals, let each player send a signal to the other, and then let them play the base game with strategies that allow the act in the base game to depend on the signal received from the other player.

If there are n possible signals, then a strategy in the cheap talk game is an n + 1-tuple:

<signal to send, act to take if receive signal 1, . . . , act to take if receive signal n>.

Thus a 2 × 2 base game with 2 signals generates a cheap talk game with 8 strategies; a 3 × 3 base game with 3 signals generates a cheap talk game with 81 strategies. If two strategies are paired in the cheap talk game, they determine acts in the base game and they receive the payoffs of their respective acts when those are paired in the base game.

Robson (1990) was the first to point out that cheap talk may have an important effect in evolutionary games. He considered a population of individuals defecting in the Prisoner's Dilemma. If there is a signal not used by this population, a mutant could invade by using this signal as a "secret handshake." Mutants would defect against the natives and cooperate with each other. They would then do better than natives and would be able to invade. Without cheap talk, a population of defectors in the Prisoner's Dilemma would be evolutionarily stable. With cheap talk this is no longer true.

This is not to say that cheap talk establishes cooperation in the Prisoner's Dilemma. Mutants who fake the secret handshake and then defect can invade a population of the previous kind of mutants. And then if there is still an unused message, it can be used by a third round of mutants as a secret handshake. It seems that the whole story may be fairly complex.

But even if all signals are used and all strategies defect, the state—although it is an equilibrium—is not evolutionarily stable. It is a mistake to assume that cheap talk has no effect.

4. Cheap Talk in a Stag Hunt

In a note provocatively titled "Nash Equilibria are not Self-Enforcing," Robert Aumann argues that cheap talk cannot be effective in the following game:

Aumann's Stag Hunt		
	c	d
(Stag)c	9,9	0,8
(Hare)d	8,0	7,7

In this base game, there are two pure strategy Nash equilibria, cc and dd. The first is Pareto-dominant and the second is safer (risk-dominant).

Aumann points out that no matter which act a payer intends to do, he has an interest in leading the other player to believe that he will do c. If the other so believes, she will do c, which yields the first player a greater payoff. One can think of c as hunting stag and d as hunting hare, where diverting the other player to hunting stag increases a hare hunter's chances of getting the hare. Then both stag hunting types and hare hunting types will wish the other player to believe that they are stag hunters.

Aumann concludes that all types of players will send the message, "I am a stag hunter" and consequently that these messages convey no information. In this game, unlike the sender–receiver games, we have a principled argument for the ineffectiveness of cheap talk.

The argument, however, is framed in a context different from the evolutionary one we have been considering. Aumann is working within the theory of rational choice, and furthermore he is assuming that the

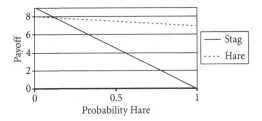

Figure 16.1 Aumann's Stag Hunt

signals have a pre-existing meaning, so that the players know which signal says "I am a stag hunter." Does the force of the argument carry over to evolutionary dynamics? We will compare the base game and the resulting cheap-talk game with two signals.

In the base Stag Hunt game there are three Nash equilibria, both hunt stag, both hunt hare, and a mixed equilibrium. As an evolutionary game we have two evolutionarily stable strategies, hunt stag and hunt hare. The polymorphic state of the population that corresponds to the mixed equilibrium is not evolutionarily stable, and it is dynamically unstable in the replicator dynamics. The dynamical phase portrait is very simple and is shown in Figure 16.1.

The state of the system is specified by the proportion of the population hunting stag. If pr(Stag) < 7/8 the replicator dynamics carries stag hunting to fixation. If pr(stag) > 7/8 replicator dynamics carries hare hunting to equilibrium. If pr(stag) = 7/8 we are at an unstable equilibrium of the replicator dynamics.

Now we embed this in a cheap talk game with 2 signals. A strategy now specifies which signal to send, what act to do if signal 1 is received, and what act to do if signal 2 is received. There are 8 strategies in this game. What is the equilibrium structure? First, we must notice that some states where everyone hunts hare are unstable equilibria. For instance, if the entire population has the strategy: "Send signal 1 and hunt hare no matter what signal you receive," then a mutant could invade using the unused signal as a secret handshake. That is, the mutant strategy: "Send signal 2 and hunt stag if you receive signal 2, but if you receive signal 1 hunt hare" would hunt hare with the natives and hunt stag with its own kind, and would thus do strictly better than the natives. The replicator dynamics would carry the mutants to fixation. Next, neither a population of hare hunters that sends both messages nor a population of stag

hunters is at an evolutionarily stable state. The proportions of those who
hunt hare and send message 1 to those who hunt hare and send message
2 could change with no payoff penalty. Likewise with the case where all
hunt stag. These states are stable in a weaker sense. They are said to be
neutrally stable, rather than evolutionarily stable. (Under the replicator
dynamics they are dynamically stable, but not asymptotically stable.)

There is, however, an evolutionarily stable state in the cheap talk
game. It is an entirely new equilibrium, which has been created by the
signals. This is a state of the population in which half the population has
each of the strategies:

<1, Hare, Stag>
<2, Stag, Hare>

The first strategy sends signal 1, hunts hare if it receives signal 1 and
hunts stag if it receives signal 2. The second sends signal 2, hunts stag if it
receives signal 1 and hare if it receives signal 2. These strategies cooperate
with each other, but not with themselves! Notice that in a population that
has only these two strategies, the replicator dynamics must drive them to
the 50/50 equilibrium. If there are more who play the first strategy, the
second gets a greater average payoff; if there are more of the second, the
first get a greater average payoff.[3]

One can check that this state is evolutionarily stable. Any mutant
would do strictly worse than the natives and would be driven to extinction
by the replicator dynamics. It is also true that this is the *only* evolutionarily
stable state in this game (see Schlag 1993; Banerjee and Weibull 2000). We
are led to wonder whether this is just a curiosity, or whether this new
equilibrium plays a significant role in the evolutionary dynamics.

Kreps' Stag Hunt		
	c	d
(Stag)c	100,100	0,8
(Hare)d	8,0	7,7

[3] This explains why there is no stable equilibrium among the strategies <1,Stag, Hare>
and <2,Hare,Stag>, which cooperate with themselves, but not with others. If one were more
numerous, it would get a greater payoff, and replicator dynamics would drive it to fixation.

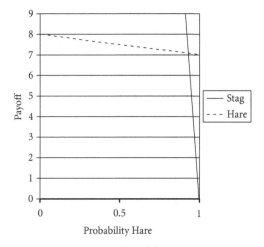

Figure 16.2 Kreps' Stag Hunt

Dynamical questions are also raised by another aspect of Aumann's paper. David Kreps asked Aumann if he would give the same analysis of the ineffectiveness of cheap talk signaling in the following stag hunt game:

Kreps' Stag Hunt		
	c	d
(Stag)c	100,100	0,8
(Hare)d	8,0	7,7

Aumann reports: "The question had us stumped for a while. But actually, the answer is straightforward: Indeed, (c,c) is not self-enforcing, even here. It does look better than in Figure 16.1 [Aumann's Stag Hunt]; not because an agreement to play it is self-enforcing, but because it will almost surely be played even without an agreement. An agreement to play it does not improve its chances further. As before, both players would sign the agreement gladly, whether or not they keep it; it therefore conveys no information." Aumann is certainly correct in that in Kreps' Stag Hunt, hunting Stag is an attractive choice even without communication. In the evolutionary dynamics, if most of the population are not initially hunting stag, hare hunting will be carried to equilibrium. The dynamical picture of the base game is shown in Figure 16.2.

But it is also true that in Aumann's Stag Hunt the odds are rigged for hunting hare. So we might want to also consider a "neutral" stag hunt, where hunting stag and hunting hare are equally attractive:

Neutral Stag Hunt		
	c	d
(Stag)c	15,15	0,8
(Hare)d	8,0	7,7

The evolutionary equilibrium analysis for Aumann's Stag Hunt holds good for all these Stag Hunts. The states in which everyone hunts stag are neutrally stable but not evolutionarily stable. The states in which everyone hunts hare are either unstable (if there is an unused message) or neutrally stable, but they are not evolutionarily stable. The unique evolutionarily stable state is again the polymorphism:

<1, Hare, Stag> 50%
<2, Stag, Hare> 50%

It is evident that in this polymorphism, the signals are certainly carrying information. This information allows individuals to always coordinate on an equilibrium of the base game. Ignoring the signals, the frequency of response types [<∗, Hare, Stag> or <∗, Stag, Hare>;] in the population is 50 percent. The signal sent identifies the response type of the sender with relative frequency of 100 percent. We should be especially interested to see whether evolutionary dynamics should lead us to expect that this type of state should be rare or frequent.

This leaves us with a number of questions about the size of basins of attraction in the signaling games. Is the size of the basin of attraction of stag hunting behavior (hare hunting behavior) increased, decreased, or unaffected by cheap talk signaling? Does the new polymorphic equilibrium created by signaling have a negligible or a significant basin of attraction?

In a Monte Carlo simulation of 10,000 trials of the neutral stag hunt game, the results were:

All hunt stag at equilibrium	75,386 trials
All hunt hare at equilibrium	13,179 trials
Polymorphism	11,435 trials

It is evident that the basin of attraction of the polymorphic evolutionarily stable state created by the signaling is not negligible. It is almost as great as the basin of attraction of the set of hare-hunting equilibrium states! *In the polymorphic equilibrium, the signals carry perfect information about the response type of the sender.* That information is utilized by the receiver, in that he performs an act that is a best response to the sender's response to the signal he himself has sent.

In an equilibrium in which everyone hunts stag, the signals carry no information about response types. Either only one signal is sent, in which case the probability of response type conditional on sending that signal is equal to the probability of that response type in the population, or both signals are sent, in which case all members of the population have the response type "hunt stag no matter which signal you receive." And in the equilibria where everyone hunts hare, both messages are always sent and everyone has the response type "hunt hare no matter which signal you receive."

Nevertheless, the signals have something to do with the basins of attraction of these equilibria. Without signaling stag hunting and hare hunting, each have basins of attraction of 50 percent of the simplex of possible population proportions. With cheap talk signals the basin of attraction of stag hunting equilibria is above 75 percent while the basin of attraction of hare hunting equilibria is below 14 percent.[4] There is something of a mystery as to how cheap talk has made this difference. We will pursue this mystery further in the next section.

5. Cheap Talk in a Bargaining Game

We will consider a simple, discrete Nash bargaining game. Each play makes a demand for a fraction of the pie. If their demands total more than 1, no bargain is struck and they get nothing. Otherwise they get what they demand. In this simplified game there are only three possible demands: 1/3, 2/3, and 1/2, which we denote by act 1, 2, and 3 respectively. The resulting evolutionary game has a unique evolutionarily stable strategy, Demand 1/2, but it also has an evolutionarily stable polymorphic state in which half of the population demands 1/3 and half of

---

[4] In Aumann's Stag Hunt the basin of attraction of stag hunting without cheap talk is only .125. With cheap talk the basin of attraction of stag hunting is increased to .149. The polymorphic Evolutionarily Stable State is also seen here, with a basin of attraction of .015.

the population demands 2/3. The polymorphic equilibrium squanders resources. Each strategy has an average payoff of 1/3. The state where all demand 1/2 and get it, is efficient. Nevertheless, if we compute the basin of attraction in the replicator dynamics of Demand 1/2, it is only about 62 percent of the 3-simplex of possible population proportions. The wasteful polymorphism has a basin of attraction of 38 percent. (Another polymorphic equilibrium, in which Demand 1/3 has probability 1/2, Demand 1/2 has probability 1/6, and Demand 2/3 has probability 1/3, is dynamically unstable and is never seen in simulations.)

What happens if we embed this game in a cheap talk game with 3 signals? A strategy in this cheap talk game is a quadruple:

<Signal to send, Demand if I receive signal 1, Demand if signal 2, Demand if signal 3>

There are now 81 strategies.

If we run a Monte Carlo simulation, sampling from the uniform distribution on the 81-simplex, and let the population evolve for a long enough time, the counterparts of the old 2/3–1/3 polymorphism in the base bargaining games are not seen. Short simulations (20,000 generations) may appear to be headed for such a state, as in Example 1.

Example 1:

pr <2112> = 0.165205
pr <2122> = 0.012758
pr <2212> = 0.276200
pr <2222> = 0.235058
pr <2312> = 0.053592
pr <2322> = 0.248208

All strategies send signal 2. Approximately half the population demands 1/3 and approximately half demands 2/3. But it is not exactly half, and the numbers do not quite sum to one because very small numbers are not printed out. This simulation (20,000 generations) has not run long enough. A population of just these strategies and with half demanding 1/3 and half demanding 2/3 would not be evolutionarily stable because it would be invadable by strategy <3113>. Message 3 functions as a "secret handshake." This strategy would get an average payoff of 1/3 when playing against the natives and 1/2 when playing against itself. Indeed, when we run the simulation longer the secret handshake strategy has time to grow, and this sort of result is never seen.

As in the previous section, the signals create new evolutionarily stable polymorphic equilibria, such as those in Examples 2 and 3:

Example 2:

pr <1132> = 0.2000000
pr <1232> = 0.2000000
pr <2331> = 0.4000000
pr <3121> = 0.1000000
pr <3122> = 0.1000000

Example 3:

pr <1312> = 0.250000
pr <2213> = 0.250000
pr <2223> = 0.250000
pr <3133> = 0.250000

These appear in simulations where the population was allowed to evolve for 1,000,000 generations.

We can gain some insight into these polymorphisms if we notice that the population can be partitioned into three subpopulations according to the signal sent. Since a player conditions her strategy on the signal received, a player uses a strategy of the base game against each of these subpopulations, and the choice of strategies against different subpopulations are logically independent.

Consider Example 2. The first thing to look at is the interaction of subpopulations with themselves. Considered alone, the subpopulation that sends signal 1 is in the 1/3–2/3 polymorphic evolutionary equilibrium of the base game. So is the subpopulation that sends signal 3. The subpopulation that sends signal 2 is in the All 1/2 evolutionary equilibrium of the base game.

The subpopulations not only play themselves. They also play each other. Notice that when two subpopulations meet they play a pure strategy Nash equilibrium of the base game. When signal 1 senders meet signal 2 senders they both demand 1/2; when signal 1 senders meet signal 3 senders the former demand 2/3 and the latter demand 1/3; when signal 2 senders meet signal 3 senders the former demand 1/3 and the latter demand 2/3. These are all strict equilibria and are stable in two-population replicator dynamics.

The three subpopulations can be thought of as playing a higher level game with payoff matrix:

Example 2′

	Sig1	Sig2	Sig3
Sig1	1/3	1/2	2/3
Sig2	1/2	1/2	1/3
Sig3	1/3	2/3	1/3

Considered as a one-population evolutionary game, this has a unique interior attracting equilibrium at $\Pr(\text{Sig } 1) = .4$, $\Pr(\text{Sig } 2) = .4$, $\Pr(\text{Sig } 3) = .2$, which are their values in Example 2.

Example 3 has a similar analysis. Here the 1/3–2/3 polymorphism occurs in the subpopulation that sends signal 2, while the other subpopulations always demand 1/2 when they meet themselves. The higher level game between subpopulations has the payoff matrix:

Example 3′

	Sig 1	Sig 2	Sig 3
Sig 1	1/2	1/3	2/3
Sig 2	2/3	1/3	1/2
Sig 2	1/3	1/2	1/2

Considered as a one-population evolutionary game, this has a unique interior attracting equilibrium at $\Pr(\text{Sig } 1) = .25$, $\Pr(\text{Sig } 2) = .50$, $\Pr(\text{Sig } 3) = .25$, as we observe in Example 3.

The foregoing modular analysis depends for its validity on the fact that all the equilibria at various levels of the story are structurally stable dynamical attractors.

These polymorphisms achieve a payoff much better than that of the 1/3–2/3 polymorphism in the base game. In the 1/3–2/3 polymorphism, each strategy has an average payoff of 1/3. In Examples 2 and 3, the average payoffs to a strategy are $0.466666\ldots$ and $0.458333\ldots$ respectively. The remaining inefficiencies are entirely due to the polymorphisms involved in subpopulations of those who send a specified signal meeting themselves.

Although these new polymorphisms are fascinating, they play a minor role in the overall evolutionary dynamics of the bargaining game. In fact, more than 98 percent of the 81-simplex of the cheap talk game evolves to one of the equilibria where all demand 1/2. The result is robust to the deletion or addition of a signal. If we run the dynamics with 2 signals or 4 signals, the resulting simulation still leads to all demanding 1/2 more than 98 percent of the time. The little mystery that we were left with at the end of the last section has become a bigger mystery.

Existing theorems about cheap talk do not apply (see Wärneryd 1991, 1993; Blume et al. 1993; Schlag 1993, 1994; Sobel 1993; Kim and Sobel 1995; Bhaskar 1998). The simulation is set up so that each run begins in the interior of the 81-simplex. That is to say that each strategy has some positive, possibly small, probability. In particular, there are no unused messages. The game is not a game of common interest. It is in the common interest of the players to make compatible demands, but within these bounds there is opposition of interests. The best result is to demand 2/3 while the other player demands 1/3.

Signals have no preexisting meaning. Does meaning evolve? Consider some of the final states of trials in which the population ends up in states where everyone demands 1/2:

Example 4 (1,000,000 generations):

pr <2131> = 0.189991
pr <2132> = 0.015224
pr <2133> = 0.037131
pr <2231> = 0.245191
pr <2232> = 0.048024
pr <2233> = 0.021732
pr <2331> = 0.128748
pr <2332> = 0.175341
pr <2333> = 0.138617

Each strategy sends signal 2; each strategy demands half upon receipt of signal 2. But concerning what would be done upon receipt of the unused signals 1 and 3, all possibilities are represented in the population.

Example 5 (1,000,000) generations:

pr <1333> = 0.770017
pr <2333> = 0.057976
pr <3333> = 0.172008

Here all messages are sent, each strategy ignores message sent and simply demands half. In between these two extremes we find all the sorts of intermediate cases.

It is clear that in our setting we do not have anything like the spontaneous generation of meaning that we see in Lewis sender–receiver games. The action of signals here is more subtle. Suppose we shift our attention from the strong notion of meaning that we get from a signaling system equilibrium in a sender–receiver game to the weaker notion of information. A signal carries information about a player if the probability that the player is in a certain category given that he sent the signal is different from the probability that he is in the category simply given that he is in the population. At equilibria where players all demand half, signals cannot carry information about the player's acts. Conceivably, signals could carry some information about a player's response type at equilibrium, where response type consists of the last three coordinates of a player's strategy. But in Examples 4 and 5 this cannot be true, because in Example 4 there is only one signal sent and in Example 5 there is only one response type.

Perhaps the place to look for information in the signals is not at the end of the process, but at the beginning and middle of evolution. For the signals to carry no information about response types at a randomly chosen vector of population proportions would take a kind of a miracle. At almost all states in the simplex of population proportions there is some information about response types in the signals. There is information present—so to speak, by accident—at the beginning of evolution in our simulations. Any information about a response type could be exploited by the right kind of strategy. And the right kind of strategy is present in the population because all types are present. Strategies that exploit information present in signals will grow faster than other types. Thus, there will be an interplay between "accidental" information and the replicator dynamics.

To investigate this idea we need a quantification of the average amount of information in a signal. We use the Kullback–Leibler (K–L) discrimination information between the probability measure generated by the population proportions and that generated by conditioning on the signal.

Let P denote the proportion of the population and p_i denote the proportion of the sub-population that sends message i. Recall that a response type consists of a triple:

<Act if message 1, Act if Message 2, Act if Message 3>.

There are 3 signals which we denote as S_i and 27 response types, which we denote as T_k. Then the K–L information in message i is:

$$\sum_k (p_i[T_k]\log(p_i[T_k]/P[T_k])).$$

The average amount of information in the signals in the population is gotten by averaging over signals:[5]

$$\sum_i P[S_i]\sum_k (p_i[T_k]\log(p_i[T_k]/P[T_k]))$$

This is identical to the information provided by an experiment, as defined by Lindley (1956), where looking at a signal is thought of as an experiment and the particular signal seen is the experimental outcome.

For example, consider the polymorphic equilibrium in the stag hunt game:

<1, Hare, Stag> 50%
<2, Stag, Hare> 50%

There are two response types present in the population and two signals. In the probabilities conditional on signal 1, there is only one response type. The information in signal 1 is just:

$$1 \log(1/(1/2)) = \log(2) = .693 \text{ (natural logarithm).}$$

The information in signal 2 is likewise log(2), as in the average information in the signals in the population. In Example 4 of this section, the average information in the population is the information in signal 2, which is always sent. This is zero, since for all response types, k, in the population $p_k = P$. In Example 5 of this section there is only one response type in the population, so the information in each message is zero. It is, however, possible for there to be positive information in an equilibrium population state in which everyone demands 1/2, for instance:

<1,3,3,1> 50%
<2,3,3,2> 50%

Only messages 1 and 2 are sent, and the response types only differ on what they would do counterfactually, in response to receiving message 3.

---

[5] The sums are over those signals that are present in some positive proportion of the population.

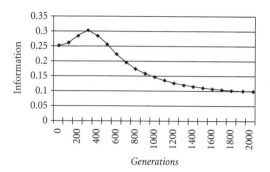

Figure 16.3 Evolution of information

The messages distinguish these response types perfectly, so they contain information about response types.

It is possible to compute the average information in the messages in a population, and to average the results over many trials in the Monte Carlo simulation used to compute basins of attraction. The results of averaging over 1000 trials are shown in Figure 16.3. (The computation uses the natural logarithm.)

There is, of course, some information present at the onset of the trials "by accident." Then the replicator dynamics leads to an increase in average information in the signals which peaks at about 300 generations. After that the average information in a signal begins a slow, steady decline.

What is the effect of this information? If there is information present in the population there are strategies present that can use it to good effect. Thus, if we look only at the behaviors in the base bargaining game—as if cheap talk were invisible—we would expect to see some departure from random pairing. That is to say, the signals should induce some correlation in the behaviors in the bargaining game. We might look to see whether Demand 1/2 behaviors are exhibited in pairs more often than at random, which would favor those types that manage to effect the correlation. We also might see if there is positive correlation between demand 2/3 and demand 1/3 behaviors. And we should be interested in whether negative correlation develops between 1/2 and 2/3 players and between pairs of 2/3 players. There would be something to be said for all the foregoing if the behaviors were fixed and self-interested. But they are not, and the evolution of behaviors is a complex product of the evolution

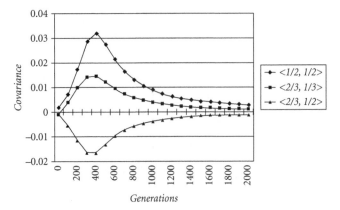

Figure 16.4 Evolution of correlation I

of signal and response. How the dynamics of all this will work out is difficult to see a priori. So we again resort to simulation. Here we calculate the covariance of the indicator variables for behaviors.

That is to say, two individual types from the 81 types in the signaling game are picked according to the current population proportions. A pair of types determines the behaviors in the bargaining game. There are indicator variables for "First member demands 2/3, 1/3, 1/2," and for "Second member demands 2/3, 1/3, 1/2." Then:

$$\text{COV}(i,j) = \text{Pr}(i,j) - \text{Pr}(i)(\text{Pr}(j))$$

where Pr(i,j) is the probability that the first demands i and the second demands j.[6] Demand 1/2 behaviors would "like" to be paired with themselves; Demand 2/3 behaviors would "like" to be paired with Demand 1/3 behaviors; Demand 1/2 behaviors and Demand 2/3 behaviors would "like" to avoid one another. Figure 16.4 shows that evolution complies to some extent. (It shows the results of simulations averaged over 1000 trials.)

The interaction between the information in the signals and the evolutionary dynamics generates a positive correlation between the compatible demands (1/2, 1/2) and (2/3, 1/3). It generates a negative correlation

---

[6] Pr(i,j) = Pr(j,i) since the draws of types are independent so Cov(i,j) = Cov(j,i).

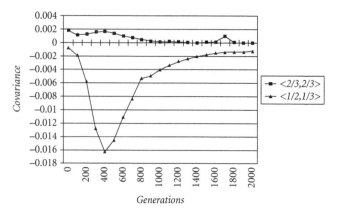

Figure 16.5 Evolution of correlation II

between the incompatible demands(1/2, 2/3). In each of these cases, the absolute magnitude of the covariance peaks at about 400 generations.

At its peak, Cov(1/2, 1/2) is above 3 percent while Cov(2/3, 1/3) is less than half of that value.

One should not be hasty, however, in generalizing the compliance of the correlation generated by evolution with what one might expect a strategist to desire. Demand 1/2 behaviors would not at all mind meeting compatible Demand 1/3 behaviors. And Demand 2/3 behaviors would "like" to avoid meeting themselves. But here evolution fails to deliver correlations in the desired directions (as shown in Figure 16.5).

Cov(1/2, 1/3) is a negative at 400 generations as Cov(2/3, 1/3) is positive. And evolution does not give Demand 2/3 behavior the edge in avoiding itself that one might expect. Demand 2/3 behaviors do not effectively use signals to avoid each other. Cov(2/3, 2/3) is mostly positive but near zero. It is evident from Figure 16.5 that thinking about what behaviors in the embedded game would "like to happen" is not a reliable guide to the working of the evolutionary dynamics in the cheap talk game.

The evolution of correlation can be put into perspective by comparing it to the evolution of bargaining behaviors, with the probabilities averaged over the same 1000 trials. This is shown in Figure 16.6.

Demand 1/2 behaviors rapidly take over the population. Demand 2/3 behaviors die out even more quickly than Demand 1/3 behaviors, notwithstanding the positive correlation between the two.

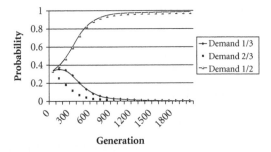

Figure 16.6 Evolution of bargaining behaviors

At 400 generations Demand 1/2 has, on average, taken over 73 percent of the population; at 1.000 generations it has, on average 97 percent of the population. At 2,000 generations, positive and negative covariance have all but vanished, Demand 1/2 behaviors have taken over 99 percent of the population, and what information remains in the signals is of little moment. In the limit, the information left in the signals, if any, is counterfactual. It only discriminates between response types that differ in their responses to signals that are not sent.

It is clear that the interesting action is taking place at about 400 generations. This suggests that we might look at individual orbits at 400 generations and then at (or close to) equilibrium. Here is an example to illustrate the complexity of the interaction of signaling strategies. It is of a printout of all strategies with more than 1 percent of the population at 400 generations and at 10,000 generations of a single run, followed by their population proportion and average payoff. At 400 generations, we have:

```
<1 3 1 1> 0.011134 U = 0.386543
<1 3 1 3> 0.044419 U = 0.436378
<1 3 2 3> 0.012546 U = 0.398289
<1 3 3 3> 0.203603 U = 0.473749
<2 1 3 3> 0.026870 U = 0.372952
<2 3 1 1> 0.52620 U = 0.385357
<2 3 1 2> 0.014853 U = 0.369061
<2 3 1 3> 0.025405 U = 0.425317
<2 3 2 3> 0.054065 U = 0.416539
<3 1 1 3> 0.022481 U = 0.354270
<3 3 1 1> 0.016618 U = 0.3954966
<3 3 1 2> 0.046882 U = 0.396710
```

<3 3 1 3> 0.050360 U = 0.416904
<3 3 2 2> 0.010570 U = 0.366388
<3 3 2 3> 0.011266 U = 0.386582
<3 3 3 1> 0.124853 U = 0.419688
<3 3 3 3> 0.010625 U = 0.440625

At 10,000 generations, we have:

<1 3 3 3> 0.973050 U = 0.500000

(with the remaining population proportions spread other the other 80 strategies).

It is evident that there is more going on here than the secret handshake. Handshake strategies are present. For example, the first strategy listed, <1 3 1 1> sends signal one, demands half when paired with anyone who sends signal one, and plays it safe by demanding 1/3 in all other cases. But there are also what might be thought of as *anti-handshake* strategies, such as <3 3 3 1>, which occupies about 12.5 percent of the population at 400 generations. This strategy demands 1/2 when it meets those who send a different signal, but plays it safe by only demanding 1/3 when it meets those who send the same signal. The anti-handshake strategy contributes to the success of the strategy that eventually takes over almost all of the population <1 3 3 3> as it eventually dies out. The same is true of the handshake strategy <1 3 1 1>.[7]

What is the cause of the dramatic difference in the magnitude of basins of attraction of the equal split produced by cheap talk (between 62 percent and 98+ percent)? We would like to have a more complete analysis, but we can say something. It appears to be due to transient information. By way of a complex interaction of signaling strategies, this transient information produces transient covariation between behaviors

---

[7] The reader may wonder, as I did, what the effect would be of removing all potential handshake strategies. I ran simulations where, in the first generation, I killed every strategy that demanded 1/2 when it met itself. Nevertheless, about 38 percent of the time the population converged to an equilibrium where each strategy demanded 1/2 of every other strategy that sent a different signal. For each signal, there were two types equally present, which implement the 1/2–2/3 polymorphism within the sub-population that sends that signal. Here is an instance:

<1 1 3 3> pr = .01667 <1 2 3 3> pr = .01667
<2 3 1 3> pr = .01667 <2 3 2 3> pr = .01667
<3 3 3 1> pr = .01667 <3 3 3 2> pr = .01667

The average payoff for each strategy is .444444.

in the base bargaining game. Much more than the "secret handshake" is involved. The type of covariation produced could not have been predicted on grounds of rational choice. Nevertheless, the net effect is to increment the fitness of strategies that demand 1/2.

6. Conclusion

Evolution matters! Analysis of the Stag Hunt and bargaining games from the viewpoint of rational choice-based game theory would not predict the phenomena discussed here. The traditional rational choice theory either makes no prediction at all, or it makes the opposite prediction from evolutionary game theory. The disparity calls for empirical testing.

Cheap talk matters! Costless signals can have a large effect on evolutionary dynamics. The signals may create entirely new equilibria and may change the stability properties of equilibria in the base game. Equilibrium analysis, however, misses important dynamic effects of signaling. Costless signals may cause large changes the relative magnitude of basins of attraction of multiple equilibria in the base game.

In the evolutionary process, information that is present "by accident" is used and amplified in the evolutionary process. Sometimes informative signaling goes to fixation. This is the case in polymorphic equilibria whose existence is created by the signals. But sometimes the information is transient and has disappeared by the time the dynamics has reached a rest state.

Transient information matters! Information, although transient, may nevertheless be important in determining the eventual outcome of the evolutionary process. Why and how this is so is a question involving the complex interaction of signaling strategies. Some insight into the nature of these interactions can be gained from simulations, but a deeper analysis would be desirable.

We can say this for certain. If costless pre-play signaling is present, then omitting it from the model on rational choice grounds may be a great error. If the signals were invisible, and an observer only saw the behaviors in the base games we have discussed, the course of evolution would not be intelligible. Mysterious correlations would come and go. The dynamics would not appear to be the replicator dynamics. Signals, even when they are cheap, have a crucial role to play in evolutionary theory.

References

Alexander, J. and B. Skyrms (1999) "Bargaining with Neighbors: Is Justice Contagious?" *Journal of Philosophy* 588–98.

Aumann, R. J. (1990) "Nash Equilibria Are Not Self-Enforcing." In *Economic Decision Making, Games, Econometrics and Optimization* ed. J. J. Gabzewicz, J.-F. Richard, and L. A. Wolsey, 201–6. Amsterdam: North Holland.

Banerjee, A. and J. Weibull (2000) "Neutrally Stable Outcomes in Cheap-Talk Coordination Games." *Games and Economic Behavior* 32: 1–24.

Bhaskar, V. (1998) "Noisy Communication and the Evolution of Cooperation." *Journal of Economic Theory* 82: 110–31.

Blume, A., Y.-G. Kim, and J. Sobel (1993) "Evolutionary Stability in Games of Communication." *Games and Economic Behavior* 5: 547–75.

Crawford, V. and J. Sobel (1982) "Strategic Information Transmission." *Econometrica* 50: 1431–51.

Grafen, A. (1990) "Biological Signals as Handicaps." *Journal of Theoretical Biology* 144: 517–46.

Kim, Y.-G. and J. Sobel (1995) "An Evolutionary Approach to Pre-play Communication." *Econometrica* 63: 1181–93.

Kullback, S. (1959) *Information Theory and Statistics*. New York: Wiley.

Kullback, S. and R. A. Leibler (1951) "On Information and Sufficiency." *Annals of Mathematical Statistics* 22: 79–86.

Lewis, D. K. (1969) *Convention: A Philosophical Study*. Oxford: Blackwell.

Lindley, D. (1956) "On a Measure of the Information Provided by an Experiment." *Annals of Mathematical Statistics* 27: 986–1005.

Nydegger, R. V. and G. Owen (1974) "Two-Person Bargaining: An Experimental Test of the Nash Axioms." *International Journal of Game Theory* 3: 239–50.

Robson, A. J. (1990) "Efficiency in Evolutionary Games: Darwin, Nash and the Secret Handshake." *Journal of Theoretical Biology* 144: 379–96.

Roth, A. and M. Malouf (1979) "Game Theoretic Models and the Role of Information in Bargaining." *Psychological Review* 86: 574–94.

Schlag, K. (1993) "Cheap Talk and Evolutionary Dynamics." Discussion Paper, Bonn University.

Schlag, K. (1994) "When Does Evolution Lead to Efficiency in Communication Games?" Discussion Paper, Bonn University.

Skyrms, B. (1996) *Evolution of the Social Contract*. New York: Cambridge University Press.

Skyrms, B. (1999) "Stability and Explanatory Significance of Some Simple Evolutionary Models." *Philosophy of Science* 67: 94–113.

Sobel, J. (1993) "Evolutionary Stability and Efficiency." *Economic Letters* 42: 301–12.

Taylor, P. and L. Jonker (1978) "Evolutionarily Stable Strategies and Game Dynamics." *Mathematical Biosciences* 40: 145–56.

Van Huyck, J., R. Batallio, S. Mathur, P. Van Huyck, and A. Ortmann (1995) "On the Origin of Convention: Evidence from Symmetric Bargaining Games." *International Journal of Game Theory* 34: 187–212.

Wärneryd, K. (1991) "Evolutionary Stability in Unanimity Games with Cheap Talk." *Economic Letters* 39: 295–300.

Wärneryd, K. (1993) "Cheap Talk, Coordination and Evolutionary Stability." *Games and Economic Behavior* 5: 532–46.

Zahavi, A. (1975) "Mate Selection—a Selection for a Handicap." *Journal of Theoretical Biology* 53: 205–14.

Zahavi, A. and A. Zahavi (1997) *The Handicap Principle.* Oxford: Oxford University Press.

17

Co-evolution of Pre-play Signaling and Cooperation

with Francesco Santos and Jorge Pacheco

A finite-population dynamic evolutionary model is presented, which shows that increasing the individual capacity of sending pre-play signals (without any pre-defined meaning), opens a route for cooperation. The population dynamics leads individuals to discriminate between different signals and react accordingly to the signals received. The proportion of time that the population spends in different states can be calculated analytically. We show that increasing the number of different signals benefits cooperative strategies, illustrating how cooperators may take profit from a diverse signaling portfolio to forecast future behaviors and avoid being cheated by defectors.

1. Introduction

In their work on the major transitions of evolution, Maynard-Smith and Szathmáry (1995) show that in each transition, there are obstacles to the emergence of cooperation that need to be overcome. Throughout most of the book these problems are modeled as the Prisoner's Dilemma games, but near the end they remark that in some cases the Stag Hunt game (the rowboat) (Skyrms 2004) may be a better model. Without precise characterizations of the payoffs involved, it may sometimes be difficult to tell which is the appropriate model. Cooperative hunting provides natural examples of Stag Hunt games (Stander 1992; Boesch 1994; Creel and Creel 1995), but let the opportunities for free riding become ample and they may degenerate into Prisoner's Dilemma games. Many popular mechanisms put forward to explain the evolution of cooperation in the Prisoner's Dilemma, such as kin selection, may transform the Prisoner's Dilemma into a Stag Hunt (Skyrms 2004). We treat both games together here.

Costless pre-play communication with signals that have no preexisting meaning (also known as cheap-talk) might not, on the face of it, be expected to do much. But Robson (1990) pointed out that such signaling can destabilize the non-cooperative equilibrium in the Prisoner's Dilemma. Suppose a mutant arises who can utilize an unused signal as a "secret handshake." The mutant sends the signal, cooperates with others who send it, and defects against the natives—who do not send it. All goes well for the invaders until another mutant arises who sends the signal and then defects. Robson's paper was followed by others using "secret handshake" arguments to establish stability properties of the efficient Nash equilibrium in games with multiple equilibria (Matsui 1991; Wärneryd 1991; Kim and Sobel 1995). In the Stag Hunt game, analysis of large population (replicator) dynamics shows that pre-play signaling can not only change the stability properties of equilibria, but can also create new equilibria, and change the relative size of basins of attraction of cooperative and non-cooperative equilibria (Skyrms 2002). Basins of attraction were there investigated by computer simulation. Here we are able to give analytical results. We present a finite population model of evolution with mutation, in which the proportion of time spent at *equilibria* can be explicitly calculated. Signaling has a strong effect. In the Stag Hunt without communication it has been long known that evolution favors the risk dominant equilibrium (Kandori et al. 1993). This is no longer true with signaling. Where the cooperative equilibrium differs from the risk dominant one, signaling can favor cooperation. Furthermore, the tilt towards cooperation (in a sense we will make precise) increases with the number of signals that are present in the population. Remarkably, this remains true even for the Prisoner's Dilemma.

2. The model

Following Skyrms (2002, 2004), if there are σ possible signals, one can define a strategy A as a vector of the form $A = \langle$ *signal, reaction to signal* $1, \ldots,$ *reaction to signal* σ $\rangle$, creating an overall set of $n_S = \sigma 2^{\sigma}$ different strategies. A simple game of cooperation without signaling can hence be viewed as a game with a single common signal, $\sigma = 1$. If people can only behave as cooperators (Cs) or defectors (Ds), when two strategies are paired the outcome of the interaction can still be described in terms of a symmetric two-player game of cooperation, with the usual payoff matrix

$$
\begin{array}{cc}
 & \begin{array}{cc} C & D \end{array} \\
\begin{array}{c} C \\ D \end{array} & \begin{pmatrix} R & S \\ T & P \end{pmatrix}
\end{array}
$$

In the following we shall study the role of signaling whenever $R > T > P > S$ (Stag Hunt) and $T > R > P > S$ (Prisoner's Dilemma).

Let us consider a finite well-mixed population of Z interacting individuals and assume that individuals revise their behaviors by social learning, implemented by means of a stochastic update rule (Nowak 2006; Traulsen et al. 2006, 2007; Sigmund 2010). At each time step an individual i with fitness Π_i (characterized here by the game payoff) will update her/his strategy by imitating a randomly chosen individual j with fitness Π_j with a probability p that increases with the increase in payoff difference between j and i (Traulsen et al. 2006, 2007). Hence, successful individuals will be imitated and the associated strategy will spread in the population. This probability may be conveniently written in terms of the so-called Fermi distribution (from statistical physics), $p = [1 + e^{-\beta[\Pi_j(k) - \Pi_i(k)]}]^{-1}$, in which β (an inverse temperature in physics) translates here into noise associated with errors in decision making (Traulsen et al. 2006). For high values of β we obtain pure imitation dynamics commonly used in cultural evolution studies, whereas for $\beta \to 0$, selection becomes so weak that evolution proceeds as random drift. It is noteworthy, however, that the following results remain robust to the adoption of other update processes, such as the Moran process, in its birth–death or death–birth variant (Nowak 2006).

We further assume that with a probability μ individuals switch to a randomly chosen strategy, freely exploring the space of possible behaviors. The ensuing analysis is largely simplified if one takes the limit of $\mu \to 0$ (so-called small-mutation limit) (Fudenberg and Imhof 2005; Imhof et al. 2005). In the absence of mutations, the end states of evolution are inevitably monomorphic, as a result of the stochastic nature of the evolutionary dynamics and update rule(s). By introducing a small probability of mutation, the population will either end up wiping out the mutant or witness the fixation of the intruder. Hence, in the small-mutation limit, the mutant will fixate or will become extinct long before the occurrence of another mutation and, for this reason, the population will spend all of its time with a maximum of two strategies present simultaneously.

Whenever two specific strategies are present in the population, say A and B, the payoff of an individual with a strategy A in a population with k As and $Z-k$ Bs can be written as $\Pi_A(k) = (k/Z)P_{A,A} + ((Z-k)/Z) P_{A,B}$, where $P_{A,A}$ ($P_{A,B}$) stands for the payoff obtained as a result of the mutual behavior (C or D) of an A strategist in a single interaction with an A (B) strategist. This allows one to describe the evolutionary dynamics of our population in terms of a reduced Markov Chain of size n_s (Fudenberg and Imhof 2005; Imhof et al. 2005), where each state represents a possible monomorphic end state of the population associated with a given strategy, and the transitions between states are defined by the fixation probabilities of a single mutant of one strategy in a population of individuals who adopt another strategy. The resulting *stationary distribution* characterizes the average time the population spends in each of these monomorphic states, and can be computed analytically. In the expression of the payoffs at the start of this paragraph, one is including self-interactions, which introduce an error, which may be sizable only in very small populations. In fact, we checked that all results below hold whether or not self-interactions are included. Inasmuch as $Z > 25$ the absence of self-interactions introduces correction below 1 percent in the stationary distributions.

Given the above assumptions, it is easy to write down the probability to change the number k of individuals with a strategy A (by $\pm$ one in each time step) in a population of $Z-k$ B-strategists:. $T^{\pm}(k) = [(Z-k)/Z] (k/Z)[1 + e^{-\beta[\Pi_A(k) \pm \Pi_B(k)]}]^{-1}$ This can be used to compute the fixation probability of a mutant with a strategy A in a population of $Z-1$ Bs. Following Ewens (2004), Karlin and Taylor (1975), Nowak et al. (2004) and Traulsen et al. (2006), it is given by, $\rho_{B,A} = (\sum_{i=0}^{Z-1} \prod_{j=1}^{i} \lambda_j)^{-1}$, where

$\lambda_j = T^-(j)/T^+(j)$. In the limit of neutral selection ($\beta \to 0$), λ_i becomes independent of the fitness values: $\rho_{BA} = 1/Z$. Considering a set $\{1, \ldots, n_S\}$ of different strategies, the fixation probabilities define n_S^2 transition probabilities of the reduced Markov Chain, with the associated transition matrix

$$M = \begin{bmatrix} 1-\eta(\rho_{1,2}+\ldots+\rho_{1,n_s}) & \eta\rho_{1,2} & \cdots & \eta\rho_{1,n_s} \\ \eta\rho_{2,1} & 1-\eta(\rho_{2,1}+\rho_{2,3}+\ldots+\rho_{1,n_s}) & \cdots & \eta\rho_{2,n_s} \\ \cdots & \cdots & \cdots & \cdots \\ \eta\rho_{n_s,1} & \cdots & \cdots & 1-\eta(\rho_{n_s,1}+\rho_{n_s,1}+\ldots+\rho_{n_s,n_s-1}) \end{bmatrix}$$

with $\eta = (n_s-1)^{-1}$ providing the appropriate normalization factor. The normalized eigenvector associated with the eigenvalue 1 of the transposed of M provides the stationary distribution described before (Fudenberg and Imhof 2005; Imhof et al. 2005). It is also noteworthy that, as the population spends most of the time in the vicinity of monomorphic states, the fraction of time the population spends in states in which individuals cooperate with its own strategy also corresponds to the fraction of time the population spends in cooperative scenarios. Consequently the stationary distribution obtained from the matrix M provides both the relative evolutionary advantage of each strategy, and also the stationary fraction of cooperative acts.

3. Results

3.1 Chatting to coordinate: the Stag Hunt game

Let us consider a population of individuals interacting through a Stag Hunt game (Skyrms 2001, 2004) with the following payoff matrix:

$$\begin{array}{cc} & \begin{array}{cc} C & D \end{array} \\ \begin{array}{c} C \\ D \end{array} & \begin{pmatrix} 1 & -0.5 \\ 0.5 & 0 \end{pmatrix} \end{array}$$

Here *Cooperation* (C) is associated with the coordinated action required to attain the highest payoff, whereas *Defection* (D) (or *non-cooperation*) counters such collective achievement, such that mutual defection leads to coordination into the least beneficial payoff. In the absence of different signals (that is, $\sigma = 1$), the evolutionary dynamics associated with this game can be described by the stationary distribution obtained via the Markov Chain with two states (Cs and Ds) and a transition matrix defined via the fixation probabilities of a single mutant. Since $\rho_{C,D} = \rho_{D,C}$ < $1/Z$, both Cs and Ds are protected against invasion and replacement by any other strategy. Moreover, for this particular payoff matrix ($T = 0.5$, $R = 1$, $P = 0$ and $S = -T$), the population can be found in a state of mutual cooperation or in a state of mutual defection with equal probabilities (see below). Hence, both cooperation and defection can be considered as *Evolutionarily Stable* in finite populations (so-called ESS_N), as defined in Nowak (2006) and Nowak et al. (2004), such that the fixation probability of any of the other strategies is always smaller than neutral fixation, that is, $1/Z$.

Let us now introduce a signaling stage before each interaction ($\sigma = 2$). Individuals can now send one of two signals ("0" or "1") without any pre-defined *meaning*, from which we obtain a Markov chain with 8 states. As all signals are considered equivalent and perfectly symmetric, all results below remain invariant if a fixed cost is assigned to each signal, apart from a trivial shift in all elements of the payoff matrix.

The results are shown in Figure 17.1. While the abundance of each signal in the population remains symmetric (Figure 17.1a), strategies that discriminate between signals prevail (Figure 17.1b). The most successful are those that react cooperatively to their own signal, promoting a positive feedback of their acts. As noted in Robson (1990), a mutant can be successful by adopting a signal that is not used in the population, and use it as a pre-play signal to ensure a posteriori coordination. This becomes clear if one analyzes the main transition probabilities between states shown in Figure 17.2. The arrows denote transition probabilities that are larger than $\rho_n = 1/Z$, that is, those that are favored by evolution, in the absence (a) and presence (b) of different signals.

While transitions between strategies of the same signal are *not* favored by evolution, the availability of different signals leads to new transitions between different *signalers* most of them favored by evolution. As shown in Figure 17.2b, these *new* transitions turn the two strategies, which are self-reinforcing and discriminative—that is, strategies that react cooperatively to their own signal and defect when facing a different signal—into the only two ESS_Ns, as no transition from such monomorphic states is favored by evolution. In addition, as shown in Figures 17.1c and 17.2b, cooperative strategies (blue circles) are favored by the transitions among different signals and prevail in the population.

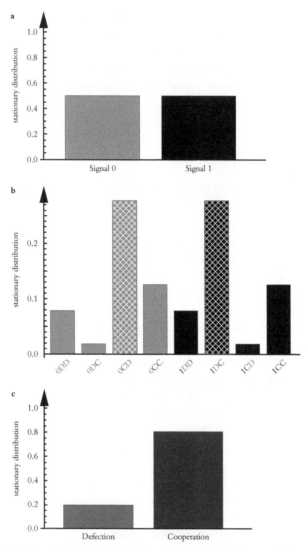

Figure 17.1 Stationary distribution for the Stag Hunt dilemma with 2 signals ("0" or "1"): (a) as expected, the evolutionary dynamics is symmetric with respect to each of the signals, (b) in the limit of rare mutations, the population spends most of the time in strategies that discriminate between signals and cooperate with their own signal (patterned bars), and (c) the accumulated frequencies of the four strategies in which individuals cooperate with each other under a monomorphic scenario, shows that, in the presence of two signals, cooperation prevails ($Z = 150$, $\beta = 0.05$, $T = 0.5$, $R = 1$, $P = 0$, $S = -0.5$).

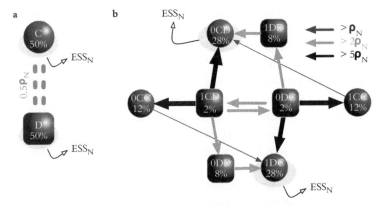

Figure 17.2 Transition probabilities favored by evolution (a) in the absence of signaling and (b) with two signals available. With pre-play signaling, strategies that (i) discriminate between signals and (ii) react cooperatively to their own signal, become the only evolutionary stable strategies in finite populations. Arrows $A \to B$ denote that a mutant B in a population of As will fixate with a probability larger than $1/Z$. Blue (red) circles (squares) stand for strategies that opt for cooperation (defection) in a monomorphic scenario. The percentage values stand for the prevalence of each strategy associated with the stationary distribution (see also Figure 17.1b). The equivalence the role played by the two signals of panel (b), results in a transition graph with a high level of symmetry. The parameters chosen are the same as in Figure 17.1. (For interpretation of the references to color in this figure legend, the reader is referred to the web version of this article.)

Without different signals, the stationary fraction of cooperative acts is dictated by the size of the basins of attraction of cooperation and defection (*risk dominance*). With signaling, this is no longer true. Whenever different signals are available, even if costless and meaningless, full cooperation will occur 80 percent of the time (for the game parameters described), contrary to 50 percent in the absence of different signals. Consequently individuals are able to assign a meaning to the signals and, as a result, modify the original basins of attraction. Conveying a signal "1" may be understood as, "*I will go for Stag, if you signal 1*" or, equivalently, the same meaning conveyed by a signal "0". In practice, however, even if in most of the stable monomorphic configurations the information portrayed by each signal is clear, the meaning of each signal emerges accidently from the stochastic nature of evolution. Meanings are therefore transient and frequency dependent, co-evolving with the

strategies present in the population and used at profit to ensure coordination (Skyrms 2010). As a result, cooperation will thrive by means of
the "secret handshake" (Robson 1990) needed for coordination, without
the necessity of keeping it "secret." In fact, *deceiving* benefits neither the
sender nor the receiver in the Stag Hunt. As a result, *true signalers* will
perform better than any mutant and emerge as ESS_N (Nowak et al. 2004),
benefiting from information content enclosed in each signal.

The reasoning above leads naturally to a different but pertinent
question: If transitions between distinct signals are favored by evolution
and, as a result, cooperation finds a window of opportunity to thrive, a
higher number of signals should enhance even more the chances of
cooperators. Since the transition from defective to cooperative states
demands the adoption of signals that are not present in the native
population, increasing the number of available signals can only ease
such process. Figure 17.3 confirms such hypothesis. As the number of
available signals increases, transitions between distinct signals become
more frequent (see Figure 17.3b). This will reinforce the flow of probabilities into cooperative strategies.

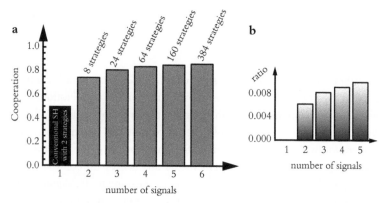

Figure 17.3 Dependence in the number of different signals: (a) fraction of time
the population spends in cooperative strategies as a function of the number of
signals available and (b) ratio between the accumulated transition probabilities of
monomorphic states with the same signal and monomorphic states with different
signals, as a function of the available number of signals. The relative change with
the number of signals shows how transitions between distinct signals become
more frequent. The parameters are the same as in Figures 17.1 and 17.2.

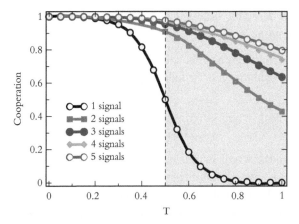

Figure 17.4 Average fraction of time spent in monomorphic cooperative states as a function of the temptation to defect T for different numbers of signals. When the number of signals increases, the original (risk-dominance) balance between cooperators and defectors in the absence of different signals (black line with open circles and vertical dashed line) is modified by a game setting in which cooperation becomes dominant for the entire range of the Stag Hunt game ($Z = 150$, $\beta = 0.1$, $S = -T$, $R = 1$, $P = 0$).

As shown in Figure 17.4—where we portray the stationary fraction of cooperative acts as a function of the temptation to defect T (with $R = 1$, $P = 0$ and $S = -T$)—the positive effect of signaling occurs independently of the particular payoffs of the Stag Hunt game matrix. In the absence of different signals, the stationary fraction of cooperative acts is defined by the ratio $\rho_{C,D}/\rho_{D,C}$ where $\rho_{C,D}(\rho_{D,C})$ is the fixation probability of a C (D) in a population of $Z - 1$ Ds (Cs). For $0 < T < 0.5$, $\rho_C > \rho_D$ whereas the opposite occurs for $0.5 < T < 1.0$. It is noteworthy, that the value of $T = 0.5$ also indicates the threshold above which defection becomes the risk-dominant behavior under the Stag Hunt, that is, the best option whenever the opponent is equally likely to play C or D (Nowak 2006). Yet, whenever the number of signals increases, this risk-dominance balance is disrupted in favor of cooperation. It is *as if* cooperation becomes the risk-dominant strategy for the entire range of parameters of the Stag Hunt game. Finally, the transition at $T = 0.5$ – associated with the change of the risk-dominant strategy of the Stag Hunt—becomes sharper with the increase of the intensity of selection β, likewise to the enhancing effect of cooperation through signaling (increasing function of β).

3.2 When deceiving becomes a profitable option: the Prisoner's Dilemma game

Let us now consider one population of individuals interacting through a Prisoner's Dilemma with the following payoff matrix:

$$
\begin{array}{c@{\;}c@{\;}c}
 & C & D \\
\begin{array}{c} C \\ D \end{array} & \left(\begin{array}{c} 1 \\ 1.5 \end{array} \right. & \left. \begin{array}{c} -0.5 \\ 0 \end{array} \right)
\end{array}
$$

Contrary to the Stag Hunt dilemma, in the Prisoner's Dilemma cooperators are always disadvantageous irrespective of the fraction of cooperators. Hence, in the absence of different signals, it is not surprising that defection emerges as the single ESS (and ESS_N) (Nowak et al. 2004; Nowak 2006). In the presence of different signals, as mutual cooperation is no longer the best possible outcome in polymorphic populations, deceiving becomes an option. Hence, signals may have the same effect as in the Stag Hunt game but, contrary to it, defectors who fake signals may end up advantageous with respect to cooperators and true signalers. Life gets harder for cooperators as now they can be betrayed at profit to the traitors.

In Figure 17.5 we depict the transition probabilities for the case with one and two available signals, following the same convention used in the case of the Stag Hunt game. For $\sigma = 2$ (Figure 17.5b), we no longer have strategies that are ESS_N, as in the case of $\sigma = 1$ (Figure 17.5a). Defection no longer works as a *sink* for all transition probabilities. Similarly to the Stag Hunt game, the transitions most favored by evolution are those between states with different signals. However, now transitions between strategies with the same signal can be favored by evolution.

Let us consider two mini-games: One (A) in which transitions take place only between strategies with the same signal (solid arrows in Figure 17.5b) and another (B) in which one considers transitions only between different signals (dashed arrows lines in Figure 17.5b). Analysis of these mini-games shows that defective strategies (red squares) are the only possible ESS_N ($1DD$ and $0DD$) in A whereas cooperative strategies (blue circles) are the only candidates for ESS_N ($0CD$ and $1DC$) in B. Thus,

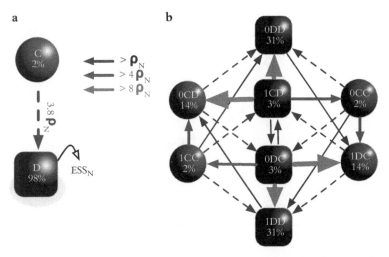

Figure 17.5 Stationary probability distribution for the Prisoner's Dilemma with (a) $\sigma = 1$ and (b) $\sigma = 2$. As before, arrows denote transitions favored by evolution and blue (red) circles (squares) denote strategies that lead to full cooperation (defection) under a monomorphic scenario. In the absence of different signals (a), defection is the only stable strategy. Yet, whenever two different signals are available (b), those strategies that opt invariably to defect are no longer stable. The overall level of cooperation results from the interplay between the *mini-game* between different signalers (pictured in solid arrows and where both cooperative strategies 0CD and 1DC are stable) and the mini-game played between strategies of the same signal (dashed arrows, where the four strategies that lead to defection are stable). Parameters: $T = 1.5$, $S = -0.5$, $R = 1$, $P = 0$, $Z = 150$, $\beta = 0.05$. (For interpretation of the references to color in this figure legend, the reader is referred to the web version of this article.)

the overall level of cooperation results from the strength and number of transitions between different signalers, from which one can infer that increasing the number of signals will favor the prevalence of cooperative strategies. This is confirmed in Figure 17.6 where we show that cooperation is promoted whenever one increases the number of signals, over a wide range of parameters of the Prisoner's Dilemma. Cooperation can emerge as a result of the arms race between (i) the exploration of new signals by cooperators (to avoid being cheated by defectors) and (ii) the search of cooperative signals by defectors (to deceive cooperators). By increasing the number of signals, cooperators

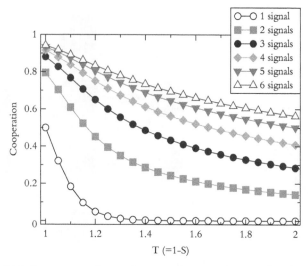

Figure 17.6 Average prevalence of cooperative strategies as a function of the temptation to defect T in the Prisoner's Dilemma for different numbers of signals. Cooperation is enhanced in the entire Prisoner's Dilemma regime whenever one increases the number of signals available. These results were obtained with same parameters as in Figure 17.5, and with $R = 1$, $P = 0$ and $S = 1 - T$.

have a larger portfolio of signals to pick from, something they learn to use to their own advantage.

4. Conclusion

We have shown analytically how pre-play signaling leads to profound changes in the evolutionary dynamics of cooperative games, favoring cooperation. As kin-discrimination, beard chromodynamics, or phenotypic diversity (Jansen and van Baalen, 2006; Traulsen and Nowak, 2007; Antal et al., 2009; Sigmund, 2009; Gardner and West, 2010), pre-play signaling represents an important correlation device, under which cooperation may prevail. Yet, here cooperation freely emerges from the co-evolution of signals and actions which are not built-in in the individual, addressing in a general framework the study of central aspects of Human evolution, from the self-organized drive towards an individual adoption of a given signaling system to the emergence of the latter (Skyrms 2002, 2010). We analyze two important metaphors of cooperation: The Stag Hunt (or coordination) game and the Prisoner's Dilemma. In coordination dilemmas, individuals willing to cooperate learn how to use the information encoded in each signal to identify other cooperators, reducing the risk of facing defection upon a cooperative act. In addition, the existence of a large number of signals enhances the tendency to cooperate, as it enlarges the portfolio of available signals that cooperators may use at profit to coordinate. Since mutual cooperation is always the best possible outcome, cooperators who are able to discriminate between their own strategy and the one of others are robust against the invasion of mutants. Consequently, the emergence of ESS_N requires that these strategies are (i) cooperative, (ii) discriminative, and (iii) self-reinforcing, that is, they cooperate with individuals who adopt the same signal.

Remarkably, the enhancement of cooperation through signaling also applies to games where deception constitutes a profitable option, and where defection is the only stable strategy, as in the Prisoner's Dilemma. In the presence of pre-play signaling, those strategies that opt invariably to defect are no longer stable in the Prisoner's Dilemma. However, the same remains true for any type of cooperative strategy. Thus, in the absence of any stable strategy, the fate of cooperation emerges from the conflict between deception by fake signaling and the development of reliable "secret handshakes" (Robson 1990). These features are strongly dependent on the number of signals available, and illustrate the advantages of a complex signaling system.

Acknowledgments

The authors acknowledge the financial support of FCT-Portugal (FCS and JMP) and the Air Force Office of Scientific Research FA9550-08-1-0389 (BS).

References

Antal, T., H. Ohtsuki, J. Wakeley, P. Taylor, and M. Nowak (2009) "Evolution of cooperation by phenotypic similarity." *Proceedings of the National Academy of Sciences* 106: 8597.

Boesch, C. (1994) "Cooperative Hunting in Wild Chimpanzees." *Animal Behaviour* 48: 653–67.

Creel, S. and N. M. Creel (1995) "Communal Hunting and Pack Size in African Wild Dogs, Lycaon-Pictus." *Animal Behaviour* 50: 1325–39.

Ewens, W. J. (2004) *Mathematical Population Genetics I.* New York: Springer.

Fudenberg, D. and L. Imhof (2005) "Imitation processes with small mutations." *Journal of EconomicTheory* 131: 251–62.

Gardner, A. and S. West (2010) "Greenbeards." *Evolution* 64: 25–38.

Imhof, L. A., D. Fundenberg, and M. A. Nowak (2005) "Evolutionary cycles of cooperation and defection." *Proceedings of the National Academy of Sciences of the United States of America* 102: 10797–800.

Jansen, V. and M. van Baalen (2006) "Altruism through beard chromodynamics." *Nature* 440: 663–6.

Kandori, M. and G. Rob Mailath (1993) "Learning, mutation and long run equilibria in games." *Econometrica* 61: 29–56.

Karlin, S. and H. M. A. Taylor (1975) *A First Course in Stochastic Processes.* London: Academic.

Kim, Y.-G. and J. Sobel (1995) "An evolutionary approach to pre-play communication." *Econometrica* 63: 1181–93.

Matsui, A., (1991) "Cheap-talk and cooperation in society." *Journal of Economic Theory* 54: 245–58.

Maynard-Smith, J. and E. Szathmáry (1995) *The Major Transitions in Evolution.* Oxford: Freeman.

Nowak, M. A. (2006) *Evolutionary Dynamics.* Harvard, MA: Belknap.

Nowak, M. A., A. Sasaki, C. Taylor, and D. Fudenberg (2004) "Emergence of cooperation and evolutionary stability in finite populations." *Nature* 428: 646–50.

Robson, A., (1990) "Efficiency in evolutionary games: Darwin, Nash, and the secret handshake." *Journal of Theoretical Biology* 144: 379–96.

Sigmund, K., (2009) "Sympathy and similarity: the evolutionary dynamics of cooperation." *Proceedings of the National Academy of Sciences* 106: 8405.

Sigmund, K., (2010) *The Calculus of Selfishness.* Princeton, NJ: Princeton University Press.

Skyrms, B., (2001) "The Stag Hunt." *Proceedings and Addresses of the American Philosophical Association* 75, 31–41.

Skyrms, B., (2002) "Signals, evolution, and the explanatory power of transient information." *Philosophy of Science* 69: 407–28.

Skyrms, B., (2004) *The Stag Hunt and the Evolution of Social Structure.* Cambridge: Cambridge University Press.

Skyrms, B., (2010) "Signals: Evolution, Learning & Information." Oxford: Oxford University Press.

Stander, P. E. (1992) "Cooperative hunting in lions—the role of the individual." *Behavioral Ecology and Sociobiology* 29: 445–54.

Traulsen, A. and M. Nowak (2007) "Chromodynamics of cooperation in finite populations." *PLoS One* 2: 270.

Traulsen, A., M. A. Nowak, and J. M. Pacheco (2006) "Stochastic dynamics of invasion and fixation." *Physical Review E* 74: 011909.

Traulsen, A., J. M. Pacheco, and M. A. Nowak (2007) "Pairwise comparison and selection temperature in evolutionary game dynamics." *Journal of Theoretical Biology* 246: 522–9.

Wärneryd, K. (1991) "Evolutionary stability in unanimity games with cheap-talk." *Economic Letters* 36: 375–8.

18

Evolution of Signaling Systems with Multiple Senders and Receivers

1. Introduction

To coordinate action, information must be transmitted, processed, and utilized to make decisions. Transmission of information requires the existence of a signaling system in which the signals that are exchanged are coordinated with the appropriate content. Signaling systems in nature range from quorum signaling in bacteria (Schauder and Bassler 2001; Kaiser 2004), through the dance of the bees (Dyer and Seeley 1991), birdcalls (Hailman et al. 1985; Gyger et al. 1987; Evans et al. 1994; Charrier and Sturdy 2005), and alarm calls in many species (Seyfarth and Cheney 1990; Green and Maegner 1998; Manser et al. 2002), up to human language.

Information processing includes filtering—that is discarding irrelevant information and passing along what is important—and integration of multiple pieces of information. Filtering systems are ubiquitous. Quorum-sensing bacteria disregard low levels of signaling molecules, and only respond to concentrations appropriate to action. The black-capped chickadee disregards calls which lack the syntactic structure that identifies a chickadee origin. Every sensory processing system of a multi-celled organism decides what information to discard and what to transmit. Integration includes computation, logical inference, and voting. Although we usually think of these operations in terms of conscious human thought, they can also be performed unconsciously by simple signaling networks. Finally, information must be used to make decisions. These decisions may have fitness consequences for the whole group, down to the level of quorum sensing in bacteria and up to alarm calls and signals indicating location and quality of food sources.

From an evolutionary perspective, these three aspects of coordination are best addressed simultaneously. They may sometimes be separable in human affairs, but elsewhere in nature it is more typical that they have co-evolved. It is possible to construct simplified models which capture essential aspects of these issues as evolutionary games.

These models may also be viewed as modules that, once evolved, may be put together to form more complex interactions. Evolutionary games may be studied from both a static and a dynamic point of view. Static analysis of equilibria reveals a lot about the structure of the interaction, and it can be carried out at a level of generality that does not commit one to a particular dynamics. But dynamic analysis sometimes reveals complexities that are not immediately apparent from the study of equilibria. Dynamic analyses may be mathematically challenging. Computer simulations are always available as a tool, but in these simple game-theoretic models analytic methods are also applicable.

We start with dyadic sender–receiver games—one sender and one receiver—and then generalize the model to multiple senders and multiple receivers. It can be shown that surprisingly sophisticated behavior can emerge from the dynamics of evolution. A full analysis, however, is non-trivial in even the simplest dyadic signaling games, and much remains to be done.

2. Classic Two-Agent Sender–Receiver Games: Equilibrium Considerations

In the basic model (Lewis 1969), there are two players, the sender and the receiver. Nature chooses a state with some probability (each state having non-zero probability of being chosen) and the sender observes the state. The sender then sends a signal to the receiver, who cannot observe the state directly but does observe the signal. The receiver then chooses an act, the outcome of which affects them both, with the payoff depending on the state. We assume at the onset that the numbers of states, signals, and acts are equal. Where this number is N, we refer to this as an $N \times N \times N$ game.

There is pure common interest between sender and receiver—they get the same payoff. There is exactly one "correct" act for each state. In the correct act–state combination they both get a payoff of one; otherwise payoff is zero. We number the states and acts so that in a play of the game, <state, signal, act> = $<s_i, m_j, a_k>$ the payoff is 1 if $i = k$, 0 otherwise.

A sender's strategy consists of a function from states to signals; a receiver's strategy consists of a function from signals to acts. Expected payoffs are determined by the probability with which nature chooses states, and the population proportions of sender's and receiver's strategies. For the purposes of evolution, individual senders and receivers are assumed to have deterministic strategies.

Signals are not endowed with any intrinsic meaning. If they are to acquire meaning, the players must somehow find their way to an equilibrium where information is transmitted. When transmission is perfect, so that the act always matches the state and the payoff is optimal, Lewis calls the equilibrium a *signaling system*. For instance, in a $3 \times 3 \times 3$ game the following combination of strategies is a Lewis signaling system equilibrium:

SENDER	RECEIVER
State 1 ⇒ Signal 3	Signal 3 ⇒ Act 1
State 2 ⇒ Signal 2	Signal 2 ⇒ Act 2
State 3 ⇒ Signal 1	Signal 1 ⇒ Act 3

as is any combination of strategies that can be gotten from this one by permutation of signals. The "meaning" of the signals is thus purely conventional, depending on the equilibrium into which the agents have settled.

There are also other equilibria in signaling games. There are *pooling equilibria*, in which the sender ignores the state, and the receiver ignores the signal. For example, suppose that state 3 is the most probable. Then the following is a pooling equilibrium:

SENDER	RECEIVER
State 1 ⇒ Signal 1	Signal 3 ⇒ Act 3
State 2 ⇒ Signal 1	Signal 2 ⇒ Act 3
State 3 ⇒ Signal 1	Signal 1 ⇒ Act 3

Since the sender conveys no information, the receiver can do no better than choose the act that pays off in the most probable state. Since the receiver ignores the signal, the sender can do no better by changing his signaling strategy.

In $N \times N \times N$ games with $N > 2$, there are also *partial pooling equilibria*, for example:

SENDER	RECEIVER
State 1 ⇒ Signal 3	Signal 3 ⇒ Act 1
State 2 ⇒ Signal 1	Signal 2 ⇒ Act 3
State 3 ⇒ Signal 1	Signal 1 ⇒ Act 3

The sender's strategy does not discriminate between states 2 and 3, and leaves signal 2 unused. Upon receiving the "ambiguous" signal, the receiver chooses optimally given the limited information that was transmitted. For larger N, there are more kinds of partial pooling equilibria, depending on which states are "pooled."

Among these equilibria, signaling systems yield optimal payoff, but this is no guarantee that one will arrive at them. They also, however, have the distinction of being *strict*, that is to say that any unilateral deviation results in a strictly worse payoff. This has the immediate consequence that in an evolutionary setting a signaling system is an *evolutionarily stable* state of the population. This is true both in a two-population evolutionary model, with a population of senders and a population of receivers and in a one-population model in which an individual is sometimes in a sender role and sometimes in a position of being a receiver.

It is also easy to see that signaling systems are the *only* evolutionarily stable states (Wärneryd 1993). In the pooling example above, a mutant sender who always sent signal 2 would do just as well as the native population. Likewise, a mutant receiver whose strategy responded differently to the signal 3 (which is never sent) would not suffer for doing so. In the partial pooling example, a mutant sender who sent signal 2 in states 2 and 3 would elicit the same receiver response, and thus would have the same payoff as the natives.

In each of these cases, the mutants do not do better than the natives. The pooling and partial pooling equilibria *are* equilibria. But the mutants do no worse, so they are not driven out. That is to say that pooling and partial pooling equilibria fail the test for evolutionary stability (Maynard Smith and Price 1973). Equilibrium analysis might then lead one to suspect that evolutionary dynamics would always (or almost always) take us to signaling systems. It is not so (Huttegger 2007a, 2007b, 2007c; Pawlowitsch 2008).

3. Dynamics

The simplest dynamic model of differential reproduction for a large population is the *replicator dynamics* (Taylor and Jonker 1978; Hofbauer and Sigmund 1998). Replicator dynamics has an alternative interpretation as a model of cultural evolution by imitation of successful strategies (Björnerstedt and Weibull 1995; Schlag 1998). It has a third interpretation as a limiting case of reinforcement learning (Beggs 2005; Hopkins and Posch 2005).

We can consider a single-population model where strategies are conditional (if sender do this, if receiver do that), or a two-population model with one population of senders and another population of receivers. Both have biological applications. A two-population model is clearly appropriate for interspecies signaling. In the case same species alarm calls, individuals are sometimes in the role of sender and sometimes that of receiver.

For a single population, let the strategies be $\{S_i\}$, let x_i be the population proportion of those who use strategy S_i and let the fitness of strategy S_i played against S_j be denoted $W(S_i|S_j)$. Then, assuming random matching, the average fitness of strategy S_i is:

$$W(S_i) = \sum_j x_j W(S_i|S_j)$$

and the average fitness of the population is:

$$W(S) = \sum_i W(S_i) x_i$$

The replicator dynamics is the system of differential equations:

$$dx_i/dt = x_i [W(S_i) - W(S)]$$

For the two-population case, let x_i be the population proportion of those who use strategy S_i in the population of senders and y_i be the population of those who use strategy R_i in the population of receivers. We again assume random matching of senders and receivers, so that:

$$W(S_i) = \sum_j y_j W(S_i|R_j) \text{ and } W(Rj) = \sum_i x_i W(R_j|S_i)$$

The average fitnesses of the sender and receiver populations respectively are:

$$W(S) = \sum_i W(S_i) x_i \text{ and } W(R) = \sum_j W(R_j) y_i$$

We consider the evolution of this two population system using bipartite replicator dynamics (Taylor and Jonker 1978; Hofbauer and Sigmund 1998):

$$dx_i/dt = x_i[W(S_i) - W(S)]$$
$$dy_j/dt = y_j[W(R_j) - W(R)]$$

In both the single population and two-population models of Lewis' signaling games, the strong common interest between sender and receiver assures *global convergence* of the replicator dynamics; all trajectories must lead to dynamic equilibria (Hofbauer and Sigmund 1998; Huttegger 2007a, 2007b).

In the case of a $2 \times 2 \times 2$ Lewis signaling game, with states equiprobable, the "hasty conclusion" from evolutionarily stability equilibrium analysis is, in fact, born out by the dynamics. Equilibria other than the signaling systems are all dynamically unstable. In both two-population and one-population models, replicator dynamics carries almost all possible population proportions to a signaling system (Huttegger 2007a, 2007b, 2007c; Hofbauer and Huttegger 2008).

But if states are not equiprobable, this is no longer so. Suppose that state 2 is much more probable than state 1. Then the receiver might just do the act that is best in state 2 and ignore the signal. And since the signal is being ignored, the sender might as well ignore the state. Consider a population in which receivers always do act 2, some senders always send signal 1 and some senders always send signal 2. Any such population is an equilibrium. We have described a set of polymorphic pooling equilibria. These equilibria are dynamically stable, even though they are not evolutionarily stable in the sense of Maynard-Smith and Price (1973). They are not strongly stable attractors in the dynamics. Rather, they are "neutrally stable" in that points near them stay near them under the action of the dynamics. But they do not attract all points near them. For instance other pooling equilibria near them are not moved at all by the dynamics. The question is whether this set of pooling equilibrium, considered as a whole, has a basin of attraction. It has been shown analytically that it does (Hofbauer and Huttegger 2008). Simulations show that the size of the basin of attraction need not be negligible. The size depends, as would be expected, on the difference in probabilities of the two states. If we were to depart from the assumption that the states have equal payoffs, it would also depend on the magnitudes of the payoffs.

Even if we keep the states equiprobable and the magnitudes of the payoffs equal, almost sure convergence to a signaling system is lost of we move from $2\times2\times2$ to $3\times3\times3$. In this game, total pooling equilibria are dynamically unstable, but there are sets of neutrally stable partial pooling equilibria like the ones discussed in the last section. It can be shown analytically that the set of partial pooling equilibria has a positive basin of attraction, and simulation shows that this basin is not negligible. (Huttegger et al. 2006).

Even with the strong common interest assumptions built into Lewis' signaling games, the emergence of signaling is not quite the sure thing that it may initially have seemed on the basis of equilibrium consider-ations. Perfect signaling systems can evolve, but it is not guaranteed that they will do so. Dynamic analysis has revealed unexpected subtleties.

There are more subtleties to explore, because the sets of suboptimal equilibria are not *structurally stable* (Guckenheimer and Holmes 1983; Skyrms 1999) Small perturbations of the dynamics can make a big differ-ence. The natural perturbation to pure differential reproduction that needs to be considered is the addition of a little mutation. We can move from the replicator dynamics to the replicator–mutator dynamics (Hadeler 1981; Hofbauer 1985). For a two-population model with uniform mutation this is:

$$dx_i/dt = x_i\,[(1-e)\,W(S_i) - W(S)] + (e/n)\,W(S)$$
$$dy_j/dt = y_j\,[(1-e)\,W(R_j) - W(R)] + (e/n)\,W(R)$$

where e is the mutation rate and n is the number of strategies. We include all possible strategies. Evolutionary dynamics is now governed by a sum of selection pressure and mutation pressure. Mutation pressure pushes towards all strategies being equiprobable, where mutation into a strategy would equal mutation out. Mutation pressure can be counterbalanced or overcome by selection pressure. But if selection pressure is weak or non-existent, mutation can cause dramatic changes in the equilibrium struc-ture of the interaction.

We can illustrate by returning to the $2\times2\times2$ signaling game, two populations, states with unequal probability. Suppose state 2 is more probable than state 1. Then, as we have seen, there is a set of pooling equilibria for the replicator dynamics. In the receiver population, the strategy of always doing act 2 (no matter what the state) goes to fixation. In the sender population there is a polymorphism between two types of sender. One sends signal 1, no matter what the state; the other sends signal 2, no matter what the state. Since there is no selection pressure between the senders' types, every such sender polymorphism is an equilibrium. Addition of *any* amount of uniform mutation leads the set of pooling equilibria to collapse to a single point at which "Always send signal 1" and "Always send signal 2" are represented with equal probability. (Hofbauer and Huttegger 2008) But all other strategies are also present in small amounts at this population state, due to the action of mutation.

The big question concerns the stability properties of this *perturbed pooling equilibrium*. Is it dynamically stable or unstable? There is no unequivocal answer. It depends on the disparity in probability between the two states (Hofbauer and Huttegger 2008). A little mutation can help the evolution of signaling systems, but does not always guarantee that they evolve.

4. Costs

Let us return to the case of $2 \times 2 \times 2$, states equiprobable, but assume that *one of the signals costs something to send, while the other is cost-free.* (We could interpret the cost-free signal as just keeping quiet.) Now there are pooling equilibria in which the sender always sends the cost-free signal and there are various proportions of receiver types.

Denoting the sender's strategies as:

Sender 1: State 1 ⇒ Signal 1, State 2 ⇒ Signal 2
Sender 2: State 1 ⇒ Signal 2, State 2 ⇒ Signal 1
Sender 3: State 1 ⇒ Signal 1, State 2 ⇒ Signal 1
Sender 4: State 1 ⇒ Signal 2, State 2 ⇒ Signal 2

and the receiver's strategies as:

Receiver 1: Signal 1 ⇒ Act 1, Signal 2 ⇒ Act 2
Receiver 2: Signal 1 ⇒ Act 2, Signal 2 ⇒ Act 1
Receiver 3: Signal 1 ⇒ Act 1, Signal 2 ⇒ Act 1
Receiver 4: Signal 1 ⇒ Act 2, Signal 2 ⇒ Act 2

Table 18.1 Payoffs if sending signal is costly

	Receiver 1	Receiver 2	Receiver 3	Receiver 4
Sender 1	2-c, 2	1-c, 1	1.5-c, 1.5	1.5-c, 1.5
Sender 2	1-c, 1	2-c, 2	1.5-c, 1.5	1.5-c, 1.5
Sender 3	1.5-2c, 1.5	1.5-2c, 1.5	1.5-2c., 1.5	1.5-2c, 1.5
Sender 4	1.5, 1.5	1.5, 1.5	**1.5, 1.5**	**1.5, 1.5**

Table 18.2 Payoffs if receiving signal is costly

	Receiver 1	Receiver 2	Receiver 3	Receiver 4
Sender 1	2-.1, 2-.1	1-.1, 1-.1	1.33-.1, 1.33	1.67-.1, 1.67
Sender 2	1-.2, 1-.1	2-.2, 2-.1	1.33-.2, 1.33	1.67-.2, 1.67
Sender 3	1.5-.3, 1.5-.1	1.5-.3, 1.5-.1	1.33-.3, 1.33	1.67-.3, 1.67
Sender 4	1.5, 1.5-.1	1.5, 1.5-.1	1.33, 1.33	**1.67, 1.67**

If signal 1 is costly, cost = $2c$, states equiprobable, and a background fitness is 1, we have the payoff matrix (sender's payoff, receiver's payoff), as shown in Table 18.1.

Sender's strategy 1 and 2 pay the cost half the time, strategy 3 all the time, and strategy 4 never. Pure Nash equilibria of the game for small c are boldfaced. (If $c > .5$ it is never worth the cost to send a signal, and the signaling system equilibria disappear.) There is also a large range of mixed strategies (corresponding to receiver polymorphisms) that are equilibria. States when receiver types are approximately equally represented and senders always send the costless signal, are such pooling equilibria.

It might also *cost the receiver something to listen*. Let us combine this with a costly message and unequal state probabilities. For example, let the probability of state 1 be 1/3, the cost of signal 1, .3, and the cost of the receiver paying attention to the signals, .1. The background fitness is 1. Then the foregoing payoff matrix changes to that displayed in Table 18.2. The *pooling equilibrium*, <sender 4, receiver 4>, where sender always sends signal 2 and receiver always does act 2, is now a *strict* Nash equilibrium of the game. Either sender or receiver who deviates does strictly worse. Thus, in both one- and two-population evolutionary

Table 18.3 Payoffs if costs are state specific

	Receiver 1	Receiver 2	Receiver 3	Receiver 4
Sender 1	**2, 2**-.1	1, 1-.1	1.33, 1.33	1.67, 1.67
Sender 2	1-.3, 1-.1	**2**-.3, **2**-.1	1.33-.3, 1.33	1.67-.3, 1.67
Sender 3	1.5-.2, 1.5-.1	1.5-.2, 1.5-.1	1.33-.2, 1.33	1.67-.2, 1.67
Sender 4	1.5-.1, 1.5-.1	1.5-.1, 1.5-.1	1.33-.1, 1.33	1.67-.1, 1.67

models, it is *evolutionarily stable* and a strong (attracting) equilibrium in the replicator dynamics.

If costs are state-specific, a rosier picture is possible (Zahavi 1975). We alter the previous example so that signal 1 is free in state 1 but costs .3 in state 2 and signal 2 is free in state 2 but costs .3 in state 1. Sender 1 now pays no penalty; sender 2 always pays .3; sender 3 pays .3 two-thirds of the time (=.2); and sender 4 pays .3 one-third of the time (=.1). This is shown in Table 18.3.

The pooling state, <Sender 4, Receiver 4>, is no longer an equilibrium at all. Given that the receiver is ignoring the message, the sender is better off switching to the costless strategy, Sender 1. If so, the receiver is better off switching to Receiver 1, yielding the optimal signaling system <Sender 1, Receiver 1>. Optimality, however, may not evolve. The *suboptimal signaling system* <Sender 2, Receiver 2>, in which the sender uses the "wrong" signals and always pays a signaling cost, is also a strict equilibrium. Both signaling systems are strong (attracting) equilibria in both one- and two-population replicator dynamic models.

5. Signaling Networks

There is no reason to limit ourselves to signaling between just two actors, one sender and one receiver. In fact, most signaling systems in nature involve multiple senders, or multiple receivers, or both. If a receiver gets signals carrying different pieces of information from different senders, the signaling system is called upon to solve some problem of information processing. Consider a toy model with two senders and one receiver:

•→•←•

Signaling complementary information

There are four states of nature, each of which occurs with non-zero probability. Two individuals are situated so as to make different incomplete observations of the state. The first sees whether it is in {S1, S2} or in {S3, S4} and the second sees whether it is in {S1, S3} or in {S2, S4}. Together they have enough information to pin down the state of nature, but separately they do not. Each sends one of two signals to a receiver who must choose one of four acts. Let's say the first send chooses "red" or "green" and the second chooses "blue" or "yellow." The payoffs favor cooperation. Exactly one act is "right" for each of the states in that each of the individuals is reinforced just in case the "right" act for the state is chosen.

In this extended Lewis signaling game the observational situation of sender 1 is characterized by a partition of the states, O_1 = {{S1, S2}, {S3, S4}}. Her signaling strategy is a function from the elements of this partition into her set of signals, {R, G}. Likewise sender 2 in observational situation O_2 = {{S1, S3}, {S2, S4}} has a signaling strategy that maps the elements of her partition into her signal set, {B, Y}. The receiver's strategy maps pairs of signals {{R, B}, {R, Y}, {G, B}, {G, Y}} into her set of acts {A1, A2, A3, A4}.

All agents get payoff 1 just in case the receiver correctly identifies the state and does the appropriate act. Payoffs are shown in Table 18.4.
A *signaling system* equilibrium is a combination of sender and receiver strategies such that the payoff is equal to one in each state. As before, a signaling system is a *strict equilibrium* of the game, and signaling systems are the *only* strict equilibria. There are lots of pooling and partial pooling equilibria.

In an evolutionary setting, this three-player game gives rise to three-population models, two-population models, and one-population models.

Table 18.4 Payoffs with two senders and one receiver

	Act 1	Act 2	Act 3	Act 4
State 1	1,1,1	0,0,0	0,0,0	0,0,0
State 2	0,0,0	1,1,1	0,0,0	0,0,0
State 3	0,0,0	0,0,0	1,1,1	0,0,0
State 4	0,0,0	0,0,0	0,0,0	1,1,1

In a one-population model, an individual's strategy would be of the form: *If sender in observational situation O_1 have this sender's strategy, if sender in observational situation O_2 have that sender's strategy; if receiver have this strategy.* The most natural two-population model has a population of senders with different observational roles and a population of receivers. In all three evolutionary settings signaling systems are the unique evolutionarily stable states. It is no longer certain that a signaling system must evolve, but it is certain that a signaling system *can* evolve. In each of these settings a signaling system is a strongly stable (attracting) equilibrium in the replicator dynamics.

Each sender's signal conveys perfect information about her observation—about the partition of states of the world that she can see. The combination of signals has perfect information about the states of the world. Exactly one state corresponds to each combination of signals. And the receiver puts the signals together. The receiver's acts contain perfect information about the state of the world. *The signaling system simultaneously solves problems of transmission and integration of information.*

The basic model admits of interesting variations. Of course there may be more senders. And depending on the act set available to the receiver, he may draw the appropriate logical "conclusion" from the "premises" supplied by the various senders (Skyrms 2000, 2004, 2008). The senders' partitions may not be fixed by nature, but may themselves evolve in the presence of information bottlenecks (Barrett 2006, 2007a, 2007b).

Error

There is another class of multiple sender models, where the question is not one of complementary information but one of error. In the previous example, senders observed different partitions but there was no error in identifying the true element of the partition. Here we suppose that the senders all observe the same states but with some error in correctly identifying them. (An alternative, essentially equivalent, interpretation of the model would locate the errors in the transmission of the signals.)

For the simplest model, suppose that there are only two states and two acts. States are equiprobable. Three senders observe the states with error probability of 10 percent, with the errors being independent between senders and between trials. Each sender sends a message to the receiver, who must then choose one of the two acts. As before, we assume that act one pays off 1 for everyone involved in state 1 and act 2 pays off 1 for everyone in state 2. Otherwise no one gets anything.

Nature here first flips a coin to pick a state, and then picks *apparent states* to present to the three senders according to the error probabilities. A sender's strategy is a function from apparent state into the set of signals, {S1, S2}. We have a choice about how to set up the receiver's strategies. If we were to assume that the receiver could distinguish between senders, we could take the receiver's strategy to be a function from ordered triples of signals to acts. But here we assume that the receiver cannot distinguish between <S1, S2, S1>, <S1, S1, S2>, and <S1, S1, S2>. The receiver here has an observational partition and can only count signals. This might be thought of as a discrete approximation to a situation where the receiver perceives an intensity arising from many chemical signals, or the sound intensity arising from many calls. A receiver's strategy is then a function from frequencies of signal received to act.

Optimal signaling in this model consists in what we might call a *Condorcet equilibrium*. There is one signal that the senders all use for apparent state 1 and another that they all use for apparent state 2. The receiver goes with a majority vote. For instance, if the senders all send signal 2 in state 1, the receiver will do act 2 if two or more senders send signal 2 and act 1 otherwise. In our example, individuals at a Condorcet equilibrium reduce their error rate from 10 percent to under 3 percent. This can be viewed as an example of information filtering, as explained in the introduction.

Rather than thinking of evolution taking place solely in the context of this game, we might assume that senders' strategies already evolved in the context of single sender–receiver interactions. Then receivers usually get one signal, or multiple agreeing signals according to the evolved signaling system, but occasionally get disagreeing signals. Slow adaptation for mixed signals in such an environment is a simple problem of optimization.

Against these fixed sender strategies, receivers who go with the majority of senders will have the greatest fitness. Then replicator dynamics will converge to the optimal receiver strategy (Hofbauer and Sigmund 1998).

But suppose we forgo this easy route and ask whether Condorcet signaling equilibria can evolve in the context of the original four-person game. Both the sender's signals and the receiver's voting rule must co-evolve. It is still possible for efficient signaling to evolve. Condorcet equilibria are strict. Consequently they are stable attractors in evolutionary versions of this game using replicator dynamics. In fact, simulations show frequent evolution of Condorcet equilibria in the forgoing model.

Variations of the parameters of the model may well lead to the evolution of voting rules different from majority rule. This is an area open for exploration. Recent rational-choice literature on strategic voting (Austen-Smith and Banks 1996; Feddersen and Pesendorfer 1998) is a source of a rich set of models that can be transposed to an evolutionary setting.

Teamwork

It is sometimes the case that a well-placed sender knows what needs to be done, and can send messages to receivers who can act, but that no one receiver can do everything that needs to be done. The sender may be the foreman, or the commander, or the brain of an organism—the team leader. Success for all requires teamwork.

There may be one sender and multiple receivers:

$$\bullet \leftarrow \bullet \rightarrow \bullet$$

For a simple teamwork problem, we suppose that there are two receivers and one sender. The sender observes one of four equiprobable states of the world and sends one of two signals to each receiver. The receivers must each choose between two acts, and the acts must be coordinated in a way determined by the state for all to get a payoff. We take payoffs to be as in Table 18.5.

Table 18.5 Payoffs in a simple teamwork situation

	<A1, A1>	<A1, A2>	<A2, A1>	<A2, A2>
State 1	1,1,1	0,0,0	0,0,0	0,0,0
State 2	0,0,0	1,1,1	0,0,0	0,0,0
State 3	0,0,0	0,0,0	1,1,1	0,0,0
State 4	0,0,0	0,0,0	0,0,0	1,1,1

We assume that the sender can distinguish members of the team, so sender's strategy maps states into ordered pairs of signals and a receiver's strategy maps her signal into her space of acts. Here the problem to be solved is a combination of one of communication and one of coordination. It is solved in a signaling system equilibrium, in which everyone always gets payoff of one. A signaling system equilibrium is again a strict equilibrium, and the unique strict equilibrium in the game. It is a strongly stable attractor in the replicator dynamics.

The example can be varied in many ways, some more interesting than others. The two receivers can be thought of as playing a rather trivial two-person game, but the game is different in every state of the world. In a signaling system, the sender can be thought of either as conveying information about the game or about the optimal act to be done. In these trivial games, these are equivalent. The example could be varied by changing the four embedded two-person games and their effect on the payoffs to the sender.

Chains

Information can flow further than shown in the models given so far. Signalers can form chains, so that information is passed along until it reaches an endpoint at which it can be used. Consider a little signaling chain.

$$\bullet \rightarrow \bullet \rightarrow \bullet$$

There are a sender, an intermediary, and a receiver. Nature chooses one of two states with equal probability. The sender observes the state, chooses one of two signals and sends it to the intermediary, the intermediary observes the sender's signal, chooses one of her own two signals, and sends it to the receiver. The receiver observes the intermediary's signal and chooses one of two acts. If the act matches the state, sender, intermediary, and receiver all get a payoff of one, otherwise a payoff of zero.

Suppose that the set of potential signals available to the sender is {R, B}, and that available to the receiver is {G, Y}. A sender's strategy is a function from {S1, S2} into {R, B}, an intermediary's from {R, B} into {G, Y}, and a receiver's from {G, Y} into {A1, A2}. A signaling system here is a triple of strategies such that the composition of sender's strategy, intermediary's strategy, receiver's strategy, maps state 1 to act 1 and state

2 to act 2. Signaling systems are the unique strict equilibria in this game and the unique evolutionarily stable states in the corresponding one-, two-, and three-population signaling games. They are attractors in the replicator dynamics. In principle, signaling chains can evolve out of nothing.

However, simulations show that in this case evolution is very slow when compared with the other signaling games discussed so far. This may simply be a consequence of the multiplicity of coordination problems that need to be solved simultaneously. The speed with which the chain signaling system can evolve is much improved if the sender and receiver have pre-existing signaling systems. They could be the same signaling system, which would be plausible if sender and receiver were members of the same population, but the signaling systems need not be the same. Sender and receiver can have different "languages" so that the intermediary has to act as a "translator," or signal transducer. Suppose that the sender sends Red or Blue and the ultimate receiver reacts to Green or Yellow as follows:

SENDER	RECEIVER
State 1 $\Rightarrow$ R	G $\Rightarrow$ Act 2
State 2 $\Rightarrow$ B	Y $\Rightarrow$ Act 1

A successful translator must learn to receive one signal and send another, so that the chain leads to a successful outcome.

SENDER	TRANSLATOR	RECEIVER
State 1 $\Rightarrow$ R	see R $\Rightarrow$ send Y	Y $\Rightarrow$ Act 1
State 2 $\Rightarrow$ B	see B $\Rightarrow$ send G	G $\Rightarrow$ Act 2

The translator's learning problem is now really quite simple. The requisite strategy strictly dominates all alternatives. It pays off all the time, while the strategies *Always send Y* and *Always send G* pay off half the time, and the remaining possibility always leads to failure. The dominated strategies are eliminated (Hofbauer and Sigmund 1998), and the correct strategy evolves.

Dialogue

The chain model showed one way in which simple interactions could be strung together to form more complex signaling systems. Here is another. Suppose that a sender's observational partition is not fixed. The sender can choose which observation to make. That is to say, she can choose which partition of states to observe. Suppose also, that the receiver's decision problem is not fixed. Nature chooses a decision problem to present to the receiver. Different sorts of information are relevant to different decision problems. Knowing the actual element of partition A (the element that contains the actual state) may be relevant to decision problem 1, while knowing the actual element of partition B may be relevant to decision problem 2. This opens up the possibility of signaling dialogue, where information flows in two directions:

$$\bullet \longleftrightarrow \bullet$$

In the simplest sort of example, nature flips a coin and presents player 2 with one or another decision problem. Player 2 sends one of two signals to player 1. Player 1 selects one of two partitions of the state of nature to observe. Nature flips a coin and presents player 1 with the true state. Player 1 sends one of two signals to player 2. Player 2 chooses one of two acts.

Suppose that there are four states, {S1, S2, S3, S4}, with alternative partitions: P1 = {{S1, S2}, {S3, S4}}, P2 = {{S1, S3}, {S2, S4}}. The two decision problems require choices in different act sets: D1 = {A1, A2}, D2 = {A3, A4}. Payoffs for the two decision problems are shown in Table 18.6.

Table 18.6 Payoffs in a dialogue situation

	Decision 1 Act 1	Decision 1 Act 2	Decision 2 Act 3	Decision 2 Act 4
State 1	1	0	1	0
State 2	1	0	0	1
State 3	0	1	1	0
State 4	0	1	0	1

Player 2 has a signal set {R, G} and player 1 has a signal set {B, Y}. A strategy for player 2 now consists of three functions, one a sender strategy from {P1, P2} into {R, G}, one a receiver strategy from {B,Y} into {A1, A2}, one a receiver strategy from {B, Y} into {A3, A4}. In a signaling system equilibrium each player always gets a payoff of one. The possibility of dialogue introduces a plasticity of signaling that is absent in fixed sender–receiver games. Signaling systems are strict, and evolutionarily stable as before.

Signaling systems can evolve in the dialogue interaction in isolation, but simulations show this process to be very slow. As in the case of chains, evolution of a signaling system is much easier if we assume that some of its parts have evolved in less complicated interactions. Player one may already have signaling systems in place for the two different observational partitions as a consequence of evolution in simple sender–receiver interactions. If so, the evolution of dialogue only requires that the second player signal the problem and the first choose what to observe. This is no more difficult than evolution of a signaling system in the original Lewis signaling game.

6. Conclusion

We have investigated the evolution of signaling in some modest extensions of Lewis signaling games with multiple senders and receivers. This discussion has focused on one particular setting—a large (infinite) population or several large populations with random interactions between individuals. Different settings would call for different relevant dynamics. A small population with random encounters calls for a stochastic model of evolution, with either a growing population or one whose size is fixed at some carrying capacity (Shreiber 2001; Benaïm et al. 2004; Taylor et al. 2004). Pawlowitsch (2007) shows that in one kind of finite population model, efficient proto-languages are the only strategies that are *protected by selection*. Individuals might interact with neighbors in some spatial structure (Grim et al. 2002; Zollman 2005). Isolated individuals might invent signaling systems by trial and error learning in repeated interactions. (Skyrms 2004, 2008; Barrett 2004, 2007a, 2007b), which might then spread by a process of cultural evolution (Komarova and Niyogi 2004). In fact, urn models of reinforcement learning are very close to urn models of evolution in a small, growing population (Shreiber 2001; Benaïm et al. 2004). It has been recently proved that reinforcement dynamics in the simplest Lewis signaling game—$2\times2\times2$ states equiprobable—converges with probability one to a signaling system (Argiento et al. 2009). Such an analytic treatment of reinforcement learning does not yet exist for more complicated signaling interactions, but simulations tend to give results parallel to the evolutionary analysis given here. This is not entirely surprising, given the close connections between reinforcement learning and the replicator dynamics (Beggs 2005; Hopkins and Posch 2005).

Simple models such as those discussed here can be assembled into more complex and biologically interesting systems. The network topologies themselves may evolve (Bala and Goyal 2000; Skyrms and Pemantle 2000). There are all sorts of interesting variations. For instance, signaling networks may allow eavesdroppers, a case well-studied in McGregor (1995). But the main business of signaling networks is to facilitate successful collective action. The simple models studied here focus on crucial aspects of coordinated action. Information is acquired by units of the group. It is transmitted to other units and processed in various ways. Extraneous information is discarded. Various kinds of computation and inference are performed. The resulting information is used to guide group decisions that lead to coordinated action. All this can happen either with or without conscious thought. These processes are instantiated in human organizations, in the coordination of the organs and cells of a multicellular organism, and even within the cells themselves. Information flows through signaling networks at all levels of biological organization.

Acknowledgments

I would like to thank Jeffrey Barrett, Simon Huttegger, Louis Narens, Don Saari, Rory Smead, Elliott Wagner, and Kevin Zollman for many discussions. Rory Smead performed the "Taking as Vote" simulations. I would also like to thank two anonymous referees for this journal who provided many helpful suggestions for improvement.

References

Argiento, R., R. Pemantle, B. Skyrms, and S. Volkov (2009) "Learning to Signal: Analysis of a Micro-Level Reinforcement Model." *Stochastic Processes and their Applications* 119: 373–90.

Austen-Smith, D. and J. S. Banks (1996) "Information Aggregation, Rationality, and the Condorcet Jury Theorem." *American Political Science Review* 90: 34–45.

Bala, V. and S. Goyal (2000) "A Non-Cooperative Model of Network Formation." *Econometrica* 1181–229.

Barrett, J. A. (2006) "Numerical Simulations of the Lewis Signaling Game: Learning Strategies, Pooling Equilibria, and the Evolution of Grammar." Working Paper MBS06-09. Irvine: University of California.

Barrett, J. A. (2007a) "The Evolution of Coding in Signaling Games." *Theory and Decision* 67: 223–7.

Barrett, J. A. (2007b) "Dynamic Partitioning and the Conventionality of Kinds." *Philosophy of Science* 74: 527–46.

Beggs, A. (2005) "On the Convergence of Reinforcement Learning." *Journal of Economic Theory* 122: 1–36.

Benaïm, M., S. J. Shreiber, and P. Tarres (2004) "Generalized Urn Models of Evolutionary Processes." *Annals of Applied Probability* 14: 1455–78.

Björnerstedt, J. and J. Weibull (1995) "Nash Equilibrium and Evolution by Imitation." In *The Rational Foundations of Economic Behavior* ed. K. Arrow *et al.*, 155–71 New York: MacMillan.

Charrier, I. and C. B. Sturdy (2005) "Call-Based Species Recognition in the Black-Capped Chicadees." *Behavioural Processes* 70: 271–81.

Cheney, D. and R. Seyfarth (1990) *How Monkeys See the World: Inside the Mind of Another Species.* Chicago, IL: University of Chicago Press.

Dyer, F. C. and T. D. Seeley (1991) "Dance Dialects and Foraging Range in three Asian Honey Bee Species." *Behavioral Ecology and Sociobiology* 28: 227–33.

Evans, C. S., C. L. Evans, and P. Marler (1994) "On the Meaning of Alarm Calls: Functional Reference in an Avian Vocal System." *Animal Behavior* 73: 23–38.

Feddersen, T. and Pesendorfer, W. (1998) "Convicting the Innocent: The Inferiority of Unanimous Jury Verdicts under Strategic Voting." *American Political Science Review* 92: 23–35.

Green, E. and T. Maegner (1998) "Red Squirrels, *Tamiasciurus hudsonicus,* produce predator-class specific alarm calls." *Animal Behavior* 55: 511–18.

Grim, P., P. St. Denis, and T. Kokalis (2002) "Learning to Communicate: The Emergence of Signaling in Spatialized Arrays of Neural Nets." *Adaptive Behavior* 10, 45–70.

Gyger, M., P. Marler, and R. Pickert (1987) "Semantics of an Avian Alarm Call System: the Male Domestic Fowl, *Gallus Domesticus*." *Behavior* 102: 15–20.

Guckenheimer, J. and P. Holmes (1983) *Nonlinear Oscillations, Dynamical Systems, and Bifurcations of Vector Fields*. New York: Springer.

Hadeler, K. P. (1981) "Stable Polymorphisms in a Selection Model with Mutation." *SIAM Journal of Applied Mathematics* 41: 1–7.

Hailman, J., M. Ficken, and R. Ficken (1985) "The 'chick-a-dee' calls of *Parus atricapillus*." *Semiotica* 56: 191–224.

Hofbauer, J. (1985) "The Selection-Mutation Equation." *Journal of Mathematical Biology* 23: 41–53.

Hofbauer, J. and S. M. Huttegger (2008) "Feasibility of Communication in Binary Signaling Games." *Journal of Theoretical Biology* 254: 843–9.

Hofbauer, J. and K. Sigmund (1998) *Evolutionary Games and Population Dynamics*. Cambridge: Cambridge University Press.

Hopkins, E. and M. Posch (2005) "Attainability of Boundary Points under Reinforcement Learning." *Games and Economic Behavior* 53: 110–25.

Huttegger, S. (2007a) "Evolution and the Explanation of Meaning." *Philosophy of Science* 74: 1–27.

Huttegger, S. (2007b) "Evolutionary Explanations of Indicatives and Imperatives." *Erkenntnis* 66: 409–36.

Huttegger, S. (2007c) "Robustness in Signaling Games." *Philosophy of Science* 74: 839–47.

Huttegger, S., B. Skyrms, R. Smead, and K. Zollman (2006) "Evolutionary Dynamics of Lewis Signaling Games: Signaling Systems vs. Partial Pooling." *Synthese* 172: 177–91.

Kaiser, D. (2004) "Signaling in Myxobacteria." *Annual Review of Microbiology* 58: 75–98.

Komarova, N. and P. Niyogi (2004) "Optimizing the Mutual Intelligibility of Linguistic Agents in a Shared World." *Artificial Intelligence* 154: 1–42.

Lewis, D. K. (1969) *Convention*. Cambridge, MA: Harvard University Press.

McGregor, P. (2005) *Animal Communication Networks*. Cambridge: Cambridge University Press.

Manser, M., R. M. Seyfarth, and D. Cheney (2002) "Suricate Alarm Calls Signal Predator Class and Urgency." *Trends in Cognitive Science* 6: 55–7.

Maynard Smith, J. (1982) *Evolution and the Theory of Games*. Cambridge: Cambridge University Press.

Maynard Smith, J. and G. Price (1973) "The Logic of Animal Conflict." *Nature* 246: 15–18.

Pawlowitsch, C. (2007) "Finite Populations Choose an Optimal Language." *Journal of Theoretical Biology* 249: 606–16.

Pawlowitsch, C. (2008) "Why Evolution Does Not Always Lead to an Optimal Signaling System." *Games and Economic Behavior* 63: 203–26.

Schauder, S. and B. Bassler (2001) "The Languages of Bacteria." *Genes and Development* 15: 1468–80.

Seyfarth, R. M. and D. L. Cheney (1990) "The Assessment by Vervet Monkeys of Their Own and Other Species' Alarm Calls." *Animal Behaviour* 40: 754–64.

Schlag, K. (1998) "Why imitate and if so, How? A Bounded Rational Approach to Many Armed Bandits." *Journal of Economic Theory* 78: 130–56.

Shreiber, S. (2001) "Urn Models, Replicator Processes and Random Genetic Drift." *SIAM Journal of Applied Mathematics* 61: 2148–67.

Skyrms, B. (1996) *Evolution of the Social Contract*. Cambridge: Cambridge University Press.

Skyrms, B. (1999) "Stability and Explanatory Significance of Some Simple Evolutionary Models." *Philosophy of Science* 67: 94–113.

Skyrms, B. (2000) "Evolution of Inference." In *Dynamics of Human and Primate Societies*, ed. T. Kohler and G. Gumerman, 77–88. New York: Oxford University Press.

Skyrms, B. (2004) *The Stag Hunt and the Evolution of Social Structure*. Cambridge: Cambridge University Press.

Skyrms, B. (2008) "Signals." Presidential Address of the Philosophy of Science Association. *Philosophy of Science* 75: 489–500.

Skyrms, B. and Pemantle, R. (2000) "A Dynamic Model of Social Network Formation." *Proceedings of the National Academy of Sciences* 97: 9340–6.

Taga, M. E. and B. L. Bassler (2003) "Chemical Communication Among Bacteria." *Proceedings of the National Academy of Sciences* 100 Suppl.2: 14549–54.

Taylor, P. and L. Jonker (1978) "Evolutionarily Stable Strategies and Game Dynamics." *Mathematical Biosciences* 40: 145–56.

Taylor, C., D. Fudenberg, A. Sasaki, and M. Nowak (2004) *Bulletin of Mathematical Biology* 66: 1621–44.

Wärneryd, K. (1993) "Cheap Talk, Coordination, and Evolutionary Stability." *Games and Economic Behavior* 5: 532–46.

Zahavi, A. (1975) "Mate Selection–Selection for a Handicap." *Journal of Theoretical Biology* 53: 205–14.

Zollman, K. (2005) "Talking to Neighbors: The Evolution of Regional Meaning." *Philosophy of Science* 72: 69–85.

Index